Successfully Starting
in Astronomical Spectroscopy
– A Practical Guide

François Cochard

edp sciences

Cover illustration: The spectral profile of the star Be V731 Tau, which covers the entire visible domain, shows a strong emission of Halpha line (in red), as well as a weaker one of Hbeta line (in blue). The background image has been made by Olivier Garde and covers a region around Antares-star.

Printed in France

ISBN(print): 978-2-7598-2026-9 - ISBN(ebook): 978-2-7598-2248-5

Contents

Note : words underlined this way are defined in the glossary.

Acknowledgements

This book is the fruit of many interactions I've had with the community of astronomers over twenty years. This community is international varied, curious, enterprising, passionate.

In this community, some taught me (a lot), and others trusted me. There are amateurs and professionals... or even both at the same time.

In particular, I wish to thank:

- Christian Buil, who taught me so much. Thanks for all the passionate exchanges !

- Valérie Desnoux, is Christian's partner. Thanks for your nice presence, for all your diving into Visual Spec, in BeSS, and in the strategic planning.

- Coralie Neiner, astrophysicist at the Observatoire de Paris, who made us discover Be stars, and has become a friend. Thanks for the long discussions to recreate the world.

- Olivier Thizy, with whom I started the adventure of Shelyak Instruments since 2006.

Then... there are all the meetings, inextinguishable sources of energy on my "path in astronomy"... Maurice Abad, Agnès Acker, Jacques Adda, Evelyne Alecian, Luc Arnold, Mathieu Barthélémy, Paolo Berardi, Laurent Bernasconi, Lionel Birée, Katherine Blundell, Michel Boer, Michel Bonnement, Franck Boubault, Sylvain Bouley, Hubert Boussier, Christophe Boussin, Jacques Boussuge, Vincent Bouttard, David Boyd, Nathalie Bressand, Jean-Jacques Broussat, Yolande Buchet, Rémi Cabanac, Martine Castets, Claude Catala, Cyril Cavadore, Pascal Chambraud, Stéphane Charbonnel, Rémy Chirié, François Colas, Sophie Combe, Pierre Cruzalèbes, Joe Daglen, Jean-Luc Dauvergne, Raymond David, Steve Dearden, Bertrand De Batz, Robert Delmas, Joël Desbordes, Pierre Dubreuil, Martin Dubs, Dominique Ducerf, Nicolas Durand, Jim Edlin, Pierre Farissier, Stéphane Fauvaud, André Favaro, Christian Feghali, Paul Felenbok, Michèle Floquet, Patrick Fosanelli, Anne-Marie Galliano, Olivier Garde, Thierry

Garrel, Christophe Gillier, Jean-Paul Godard, Thierry Godard, Keith Graham, Joan Guarro, Patrick Guibert, Ken Harrison, Anton Heidemann, Huib Heinrichs, Christian Hennes, Anne-Marie Hubert, Ken Hudson, Jak de Jesus, Stella Kafka, Hugo Kalbermatten, Alain Klotz, François Kugel, Olivier Labreuvoir, Robin Leadbeater, Pascal Le Du, Steve Lee, Jean-Christophe Le Floch, Thierry Lemoult, Auguste Le Van Suu, Arnaud Leroy, Bernard Leroy, Jean Lilensten, Alain Lopez, Paul Luckas, Pierre Maquart, Vincent Marik, Gérard Martineau, Jean-Pierre Masviel, Stéphane Mathis, Benjamin Mauclaire, Philippe Michel, Jacques Michelet, Richard Monnerot, Romain Montaigut, Claire Moutou, Patrick Pelletier, Sandrine Perruchot, Éric Piednoël, Jean-François Pittet, Michel Pujol, Ernst Pollmann, Franck Razafimaharo, Christian Revol, André et Sylvain Rondi, Jean-Paul Roux, Jean-Pierre Rozelot, Raymond Sadin, Èric Sarazin, Jean-Pierre Sarreyan, Carl Sawicki, Mathieu Senegas, Joël Setton, Steve Shore, Alain Soutter, Jean-Noël Terry, François-Mathieu Teyssier, Bernard Trégon, Franck Valbousquet, Céline et Sébastien Vauclair, Sylvie Vauclair, Adrien Viciana, Brigitte Zanda...

A thought for my children, Julien, Marion, and Armand – your freshness is so good for me – and for Nathalie, with whom I invent my path everyday. Finally, a thought for my two sisters Marie and Cécile, and ...

for father, who gave us roots,
for mother, who gave us wings.

The Word from Mathieu Renzo, the Translator of the English Edition

There is no need to explain to amateurs why doing astronomical observations is cool. Almost everyone experiences some degree of amazement when they see a picture of the beauty up there in the sky, even those who don't care about things so distant from themselves. And being the person that reveals that beauty with their telescope is a great satisfaction. But there is more than just a challenge and an amusement in those pictures : they show us extreme (and most of the time hostile) conditions that we will never be able to reproduce on Earth. This is why astronomy is not only the first science that developed in the human history, but also a constant driver of scientific progress, always offering new puzzles to theorists.

Almost any field of modern physics stems directly from astronomy, and to explain astronomical observations we need almost every piece of modern physics available. The connection of some fields of physics with astronomy is trivial, for example Newtonian mechanics or General Relativity. But maybe not everybody knows that, for example, quantum mechanics was largely developed to explain stars, and in particular their spectra. And then nuclear physics (even if with a large and unfortunate contribution from military research), to explain why stars live so long, and so on. Even particle physics needs to fit within the big picture of cosmology.

The acquisition of spectra, that is the study of how the light entering the telescope is distributed in energy marks the transition from astronomy to astrophysics: it opens the door to go from observing to understanding. And understanding – even in a regime so far from what we might ever find on Earth – empower us. People often ask, for example, "why should we spend money for astronomy? Why not put all the money in curing cancer?" Well, one way to cure cancer is hadron therapy, which requires a good understanding of nuclear physics, that we acquired (also) looking at how stars live, evolve, and die!

Astronomy (or astrophysics, if you prefer) is also special among the sciences: we have no way to "control" what we can only observe, and therefore, except in very rare exceptions, we cannot perform experiments in the lab sense of the term. Nevertheless, by setting up your instrumentation, taking your first spectra, and asking yourself the questions that this book will suggest you, you can indeed practice the scientific method from which so much of our everyday life is based, but that so many people don't seem to understand.

One very important thing to note is that amateur astronomers did, can, and do contribute significantly to the scientific enterprise of understanding astrophysical phenomena. It doesn't necessarily take a professional, or a huge telescope: the main ingredient is just passion. Those below are just two examples that make no justice to the achievements of amateur astronomers.

Collectively, amateurs can observe every night, and provide uniform and long-term coverage of astronomical objects. For example, it is thanks to amateurs that we have more than 100 years of light curves of the dwarf nova SS-Cygni: every night someone, somewhere, observed it, even during the world wars. And this data set is still a precious gold mine to understand the physics of accretion disks.

Moreover, amateurs can, with a bit of luck, observe transient phenomena earlier than professionals. And their data, especially if spectroscopic, can be precious. For example, SN2016gkg has been discovered first by an amateur (who also took a spectrum of it!), and the prompt communication to the community has allowed for the study of the evolution of the spectrum of that explosion during the first few hours. This allowed observers (professionals and amateurs together) to probe events that happened in the last decades of the life of the star.

If you do good observations, like this book will teach you, and communicate them to other researchers, you are doing science. It doesn't take necessarily a fancy degree: again, most of it is just passion. Not being an amateur astronomer myself, I was glad to contribute to the translation of this book. I am a PhD student in theoretical and computational astrophysics, and my everyday work is far from the operation of telescopes and design of instruments. Translating this book was a very good way of refreshing this topic, and gain insight especially on the instrumental design. I hope this book will help passionate people contribute to the scientific endeavor of humanity, in ways that busy professionals competing with each other for telescope time cannot do.

Preface

With a history of over five thousand years, astronomy is one of the oldest sciences. Our remote ancestors understood the need to measure, understand and exploit the movements of the stars in the sky to develop accurate and reliable clocks and calendars, but they were also certainly fascinated by the beauty, the immensity and complexity of the universe, as we still are today.

The appeal of the majesty of the spectacle that the night sky offers is certainly the primary motivation for most amateur astronomers, and it must also be said that it is the source of many professional astronomers' vocation. But traditionally, while the amateur astronomer's aim is to make a good observation, sometimes requiring a big effort from instruments he often made or improved himself, the professional astronomer rather seeks, through observation, to understand the nature of the stars he/she observes.

It is fascinating for a professional astronomer like me to find that the search for knowledge, this passion that drives me like all my fellow astronomers to discover the laws that govern the behavior of celestial objects, is actually widely shared by the community of amateur astronomers. From the love of beauty to the love of science, is only one step.

The step that François Cochard invites us to take with this beautiful work, is targeted to guide us pleasantly and efficiently down the road of astronomical spectroscopy and its use by amateur astronomers. Spectroscopy consists in decomposing light depending on its wavelength, in a more or less precise way depending on the performance of the instrument used. The application of spectroscopy to astronomy is as old as spectroscopy itself. From the nineteenth century, physicists and astronomical visionaries like Joseph von Fraunhofer, Robert Bunsen and Gustav Kirchhoff in Germany or Jules Janssen and Henri Deslandres at the Observatory of Paris, pointed their spectrographs towards the Sun, discovering the amazing richness of the solar spectrum and began to deduce the composition and characteristics of our day star.

Spectroscopy is a formidable tool which gives us access to a wealth of information from analysis of the light from the objects being observed. It is thus possible to determine the chemical composition of a star, its speed relative to the Earth, its temperature, rotational speed, etc. Thus, by developing and perfecting this amazing tool, astronomers have learned over time to measure the speed of distant galaxies and to deduce properties of the expansion of the universe. They have also managed, by measuring with high accuracy the radial velocity of stars through spectroscopy, to detect the tiny movement imparted to these stars by the revolution of planets around them. These are the famous exoplanets which have been the subject of a fruitful twenty-year hunt.

This fantastic tool is now within the reach of amateur astronomers, as shown in this book. Technological advances that first led to the popularization of CCD detectors have now generated a range of spectrographs accessible to all, for all kinds of budgets. And so amateur astronomers can now, in addition to enjoying the beautiful images of heavenly objects, carry out these precise measurements like professional astronomers.

We are at the dawn of a golden age where amateur and professional astronomers will combine their talents and efforts to advance the understanding of the universe. Where professionals develop unique instruments around the world, ultra-sophisticated, mounted on giant telescopes located at the ends of the world or even in space, but which they can only use sparingly, amateurs, armed with their telescopes and spectrographs which are certainly less powerful but much more numerous and versatile, can make their unique contribution.

By coordinating their observations through a dialogue with the professionals, amateurs can contribute to the ongoing research programs by providing valuable data to complement those acquired on major advanced instruments. This may take the form, for example, of systematic monitoring over time of variable stars, or spectroscopic observations of large samples of stars that professionals are struggling to obtain due to the difficulty of access to large instruments.

The book you are about to read is remarkable in the sense that it makes the foundations of astronomical spectroscopy accessible to all and provides practical advice for its application. It will without doubt give you the desire to embark on this great adventure, and provides you the means to achieve it.

Claude Catala
Président de l'Observatoire de Paris

Introduction

The first time that I saw the spectrum of a star, with a DIY construction of wood and a small diffraction pattern assembled behind my telescope, I did not believe my eyes. The emotion was as intense as when I discovered the craters of the Moon in binoculars or Saturn's rings in a small telescope. I experienced a feeling of giddiness, that touched something incredible. From that moment (it was at the beginning of the 2000's), much water has gone under the bridge but the emotion remains. And I know now that it is a contagious emotion. Each time I had the occasion to make a presentation on spectroscopy, I could see that the subject made the eyes of the audience light up. "You say that with a small instrument in my backyard, I can measure speeds, temperatures, periods of rotation, and chemical compositions of stars?"

Yes – and much more.

Several Miracles

I've had the chance to participate in the development of this discipline at the heart of amateur astronomy. With hindsight, I see it as an adventure at the crossroads of several "miracles".

- First, there is astrophysics itself. The dimensions, the distances, and the masses of stars are beyond the reach of our mind. Nevertheless, their light carries information that we can decipher and comprehend. We will never visit a star, and yet we can understand how they work with astonishing precision. I like Einstein's quote: "The most incomprehensible thing about the world is that it is comprehensible".

- Then there is the tremendous work that was done by generations of pioneers, adventurers, and researchers, that give us the keys to understanding starlight today. To carry out astronomical spectroscopy, is to pay tribute to them, a journey to meet a wonderful monument of human knowledge, built stone by stone and whose construction continues today.

– Then there is the strength of the meetings, between some enthusiastic individuals - amateurs and professionals - with strong and complementary skills that have helped to achieve some crazy ideas: from prototyping, spectroscopes to creating a company like Shelyak Instruments[1] that designs, manufactures and distributes new instruments ... strong human experience.

– I gradually realized that when amateurs make observations in spectroscopy, they do not just follow the professionals by reproducing well-known decades old experiments. With the ability of observing "collectively" – i.e. in cooperation with other amateurs – we constitute a unique instrument that allows observations inaccessible to the professionals. Today, collaborations between amateurs and professionals are increasing across the world and are establishing an international community: it's only the beginning of the adventure.

Today, when we leave scientific research to "specialists confined in their laboratories" (in fact in their observatories), astronomical spectroscopy reminds us that research is above all a matter of curiosity, and is accessible to everyone.

I am convinced that this is human curiosity - the basis of scientific research – which is the reason for much of the success spectroscopy experiences today.

I've had the opportunity to see many people start spectroscopy, and I was often surprised to see that some of them were starting from scratch in observational astronomy - that is to say, they had no previous experience in handling a telescope or a CCD camera. For these people spending hours taking images of the sky was not enough motivation. However, to make temperature or speed measurements with a small instrument was enough motivation to take action and invest in equipment. I mentioned earlier my emotion when I achieved my first spectrum: part of that emotion was that I had sitting in front of me a very basic assembly, on a very modest telescope. No, really, getting a spectrum is not complicated.

Spectroscopy... Easy ?

I have spent a lot of time pushing this idea: spectroscopy is easy. It is, in some sense, even easier than deep sky imaging.

Nevertheless, I have seen many beginners start with enthusiasm...and fall onto very concrete problems, and sometimes give up. I have regularly been annoyed by the fact that observers come to us with problems that have little to do with spectroscopy: difficulties in installing some software, in guiding their telescope, in handling the CCD images or pointing to a star... Spectroscopy is easy, but you still need to handle some technical tools before starting!

---

[1] URL: http://www.shelyak.com

I have been irritated, but a little voice was telling me that these problems need to be faced when walking the road to "easy spectroscopy". I know how frustrating it can be to work on "magical observation", and be stopped by a ridiculous technical detail – as many observers, I have wasted many observing nights because of silly mistakes. Then, would spectroscopy be an activity reserved to an expert elite? It shall not be! Many elements need to be mastered, but none of them is very complicated.

After a few years within the community of observers, I can list these elements that need to be mastered for spectroscopy to be "relaxed and effective". This thought – together with the ever increasing need for education in this discipline – led me to write this book. My declared objective is therefore to allow you to start astronomical spectroscopy smoothly and in a rational way – and, if you are still hesitating, to convince you that you can produce breath taking observations with just a little effort.

You will find in the following pages some practical advice to get started. I want to emphasize this practical aspect: I do not intend to give lectures in optics or astrophysics, but I want to guide you to your very first own spectra.

The Three Components of Spectroscopy

This book is organized based on a simple observation that allowed me to understand the important differences between observers. It's not a matter of skills or experience, but rather of personal paths.

Scientific research – especially in astrophysics – relies on three main actors, which allow each other to bring innovations. These are theoreticians, instrumentalists and observers (Fig. 1). A new theory, an innovative instrument or unpublished observation can each "move the lines" of research.

One can make an analogy with amateur astronomical spectroscopy ; you will get good results if you have a sufficient background in three key but very different areas. (Fig 2)

- Astrophysics: understanding the nature of light and what it tells us about the stars (theory);

- Understand how a spectroscope works (instrumental part);

- Know how to carry out practical observations in the field (practical part)

As an amateur astronomer, it is very likely that you are already familiar with at least one of these three areas: there are those who have discovered the secrets of the sky from books, those who have stretched out in the grass on a summer evening with a pair of binoculars, and those who have built (at least partially) their instrument or written software to process data. Obviously everyone has their own story, and all the combinations are possible.

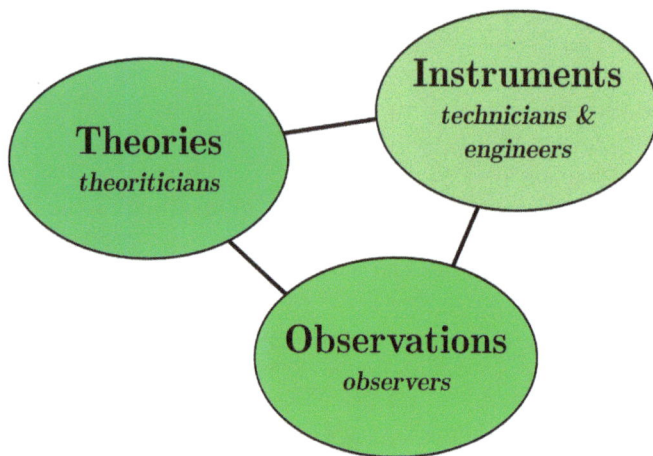

FIG. 1 – The three components of research in astrophysics.

It is therefore likely that some parts of this book may seem obvious or unnecessary. Nevertheless, I invite you to browse them quickly; there are always things to learn.

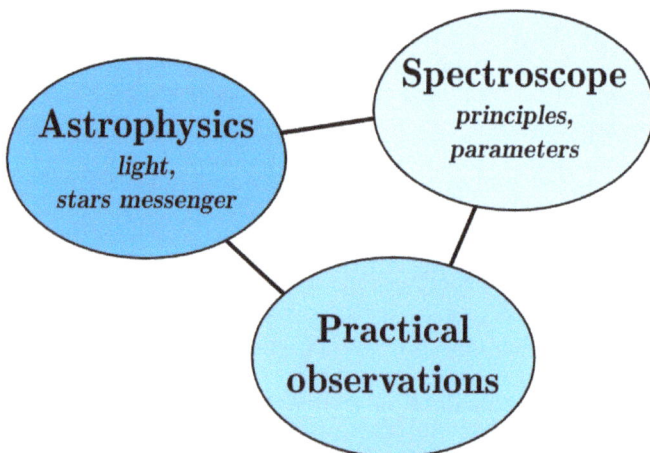

FIG. 2 – The three components of amateur astronomical spectroscopy.

Spectroscopy is a kind of "complete astronomy," similar to a complete sport that makes you work all the muscles If there is a difficulty in spectroscopy, it is hidden in the variety of areas to cover, including the actual observation in the field.

An essential key for the next step is to keep common sense constantly on standby. We are in the domain of astrophysics: there is nothing more deterministic! I know that when faced with a technical problem – especially late at night – we can exhibit very irrational behavior... however, in astronomy, we can explain a lot of things very simply.

My Background

I am a French amateur astronomer "practicing" for more than twenty years. I am not a very intensive observer – for lack of time, like many people. But I found over time, I have learned some of the essential elements that make "a good" observation.

I am member of CALA (Club of Astronomy of Lyon Ampère[2]). It is a dynamic and convivial association, with whom I was able to participate regularly in their astronomical missions or campaigns.

I am, in addition, a mechanical engineer by training, and I realized a few years ago, I could apply my knowledge of engineering to my practice of astronomy. Astronomy has the peculiarity of being both a field to stretch the limits of human thought (infinity, the place of Man in the Universe, fear of the unknown, etc.) and a very technical field (mechanics, electronics, data processing). I like to stand at this crossroads of philosophy, technology and science.

I am the co-founder along with Olivier Thizy of Shelyak Instruments[3] which develops, manufactures and markets a range of spectroscopes intended for astronomy, for both amateur and professional. In this context, I was strongly involved in the design of several instruments: Lhires III, LISA, Alpy 600, eShel, etc.

I have also actively participated in the creation of the Be star observing program (BeSS[4] and ArasBeAM[5]), in collaboration with the Paris-Meudon Observatory; this program is a real success (with over 125,000 registered spectra, including 66,000 made by amateurs at the time of writing), and is often considered a benchmark "Pro-Am" project (collaboration between professionals and amateurs).

When you practice astronomy and become serious, there comes a time when you will come into contact with the domain of professional astronomy. Before getting to know them, astrophysicists seem like demigods... they are

---

[2] URL: http://www.cala.asso.fr/
[3] URL: http://www.shelyak.com
[4] URL: http://basebe.obspm.fr/basebe/Accueil.php?flag_lang=en
[5] URL: http://arasbeam.free.fr/

actually normal human beings! I learned a lot from them (they gladly share their passion), but I also know that amateur astronomers should not have an inferiority complex: we have a rightful place to (re)take our spot in research.

I am especially motivated by the human adventure as a technical and scientific adventure, and my deepest desire is to help lots of people dreaming to enter the realm of spectroscopy to pass from dream to reality - with the underlying idea that the more observers; the more amateurs will take an active role in research.

Chapter 1

Entering the Realm of Amateur Astronomical Spectroscopy

1.1 Spectrography, Spectrometry, Spectroscopy...

I am going to quickly get out of the way a topic of recurrent debate: what's the subtle difference between spectrography, spectrometry and spectroscopy? It is only a matter of vocabulary. If I stick to the etymology, spectroscopy is the observation of spectra, spectrography is the act of recording them on a physical media, and spectrometry is the act of making measurements on them. In everyday life, the three terms are used interchangeably, and they can be considered as synonyms. In the remaining of this book, out of coherence, I have chosen to use only one of these terms – and I choose *spectroscopy*.

1.2 What Does a Spectrum Look Like?

A spectrum can take many shapes: you will sometimes see images of spectra (called 2D spectra, i.e. two dimensional). These are the raw images as they come out of our instruments (1.1).

The actual spectrum is the white line in the middle of the frame. It is the image of a star stretched horizontally, each point along the spectrum corresponds to a specific wavelength. In the following example (which is just a zoom in of the previous image of 21 Lyn), you can see some dark lines: this are absorption lines (specifically from hydrogen, in this case) (fig. 1.2).

These images can be colored (since the light they represent is colored), but more often they are in black and white. In reality the CCD detectors that we employ are not color-sensitive – and you will see later that it is much better this way. When an image is colored, it is either because they have been produced by a color-sensitive detector (like a digital camera for example), or because they have been artificially colored for the sake of display.

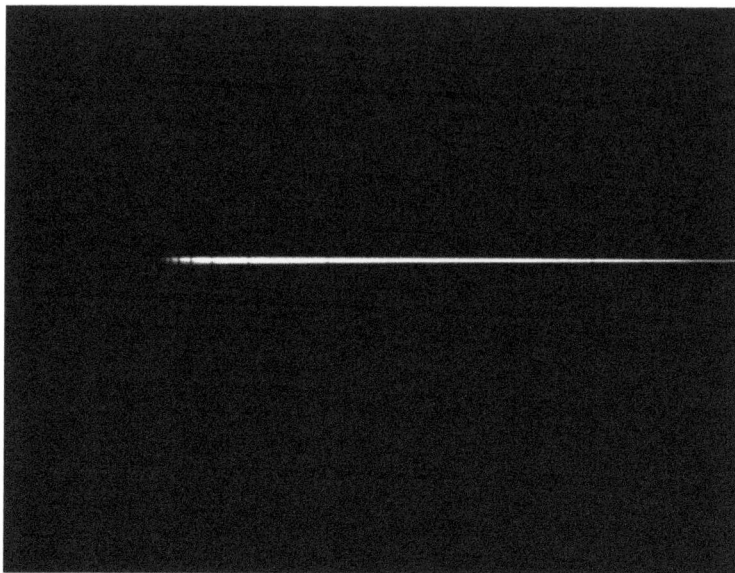

FIG. 1.1 – Raw image of the spectrum of the star 21 Lyn.

FIG. 1.2 – Magnified image of the spectrum of the star.

A stellar spectrum is generally a simple horizontal line in a frame (see figure 1.2), but when observing an extended object, its spectrum can be very broad. This is the case, for example, of a nebula, the solar spectrum, or a spectral lamp (fig. 1.3). The height of the spectrum is because the spectroscope includes a slit (positioned vertically in the images show here), and the spectrum is the dispersion of the light going through this slit.

Since the information we are interested in is the variation of the intensity of the light *along* the dispersion axis (horizontal in the figures below), we also sometimes adopt a different representation, easier to analyze by eye: it is an enlarged image of the spectrum (realized via computer processing) (fig. 1.4).

All these images are very visual and they tell a lot to a trained eye, but the most complete form of a spectrum, the one really scientifically exploitable, is a 1D spectral profile (i.e. in one dimension) with wavelength as x-axis and the corresponding intensity as the y-axis. The examples shown in fig. 1.5 give an idea of the variety of spectra that one can observe in the sky: along with the solar spectrum, you can see the spectrum of a hot star (21 Lyn, spectral type A0), a cold star (HR 5086, spectral type K5), and a carbon star (Y CVn).

FIG. 1.3 – 2D spectrum of the Sun (left) and of a calibration lamp (right)

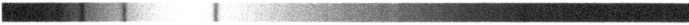

FIG. 1.4 – enlarged image of the spectrum of 21 Lyn.

1.3 The Missing Link

You have probably already read popular science books or astronomy magazines about the Big Bang, the structure and evolution of the Universe, the life and death of stars, the search of Earth-like planets beyond the Solar system (are we alone in the Universe?).

Probably you have already had the chance to lie down in a grass field on a summer night, far from the city lights, with a pair of binoculars – or with a friend with their instruments, discovering the constellations, sky rotation, some spectacular objects (lunar craters, Saturn rings, the Andromeda Galaxy, some stellar clusters...). Inevitably this leads to questions about the immensity of the sky, the distances, time – to the haze of the infinity.

What is the relationship between these two worlds, the one of books and magazines, and the one of personal experiences? How did generations of researchers manage to unravel the mechanisms driving in the Universe, just from these simple observations accessible to anyone – since you just need to lift the eyes?

Science nowadays relies on very impressive instruments – telescopes of many meters in diameter situated in inaccessible locations (or even carried on satellites or planetary spacecraft) and instruments filled of electronics and computers. The outcomes are often breathtaking images – and this might give the impression that it is thanks to the hugeness of their instruments that researchers can see what remains inaccessible to common mortals. Nothing is more wrong: the pioneers who made the founding discoveries in astronomy (many centuries ago) had way less technical capabilities than those nowadays accessible to a very ordinary amateur astronomer.

21 Lyn - Apr 18th, 2015 - F. Cochard - Alpy 600 C8 Atik314L+

Sun - Apr 15th, 2015 - F. Cochard - Alpy 600 C8 Atik314L+

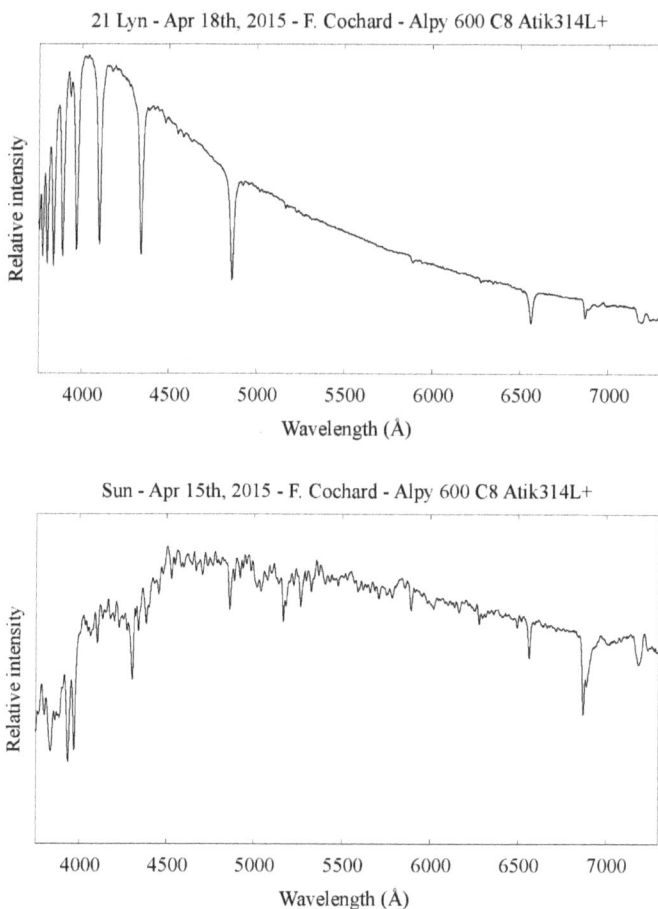

FIG. 1.5 – Several examples of spectra

No, the key for this comprehension of the universe is not in the size of the instruments, but in the nature of light itself. Our eye sees stars just as shiny dots, some times colored. Nevertheless, this light is rich with an incredible quantity of information, and our ancestors – Isaac Newton first of all – taught us little by little how to decode this information.

We had to understand that light is an electromagnetic wave, that we can stretch it as a function of its wavelength (or color). This decomposition is what we call a spectrum – and this decomposition can be realized very easily, for example by using a prism.

We had also to understand that the realization or modification of this spectrum (i.e. this light) reveals many different and complementary physical

HR5086 - Apr 18th, 2015 - F. Cochard - Alpy 600 C8 Atik314L+

Y CVn - Apr 18th, 2015 - F. Cochard - Alpy 600 C8 Atik314L+

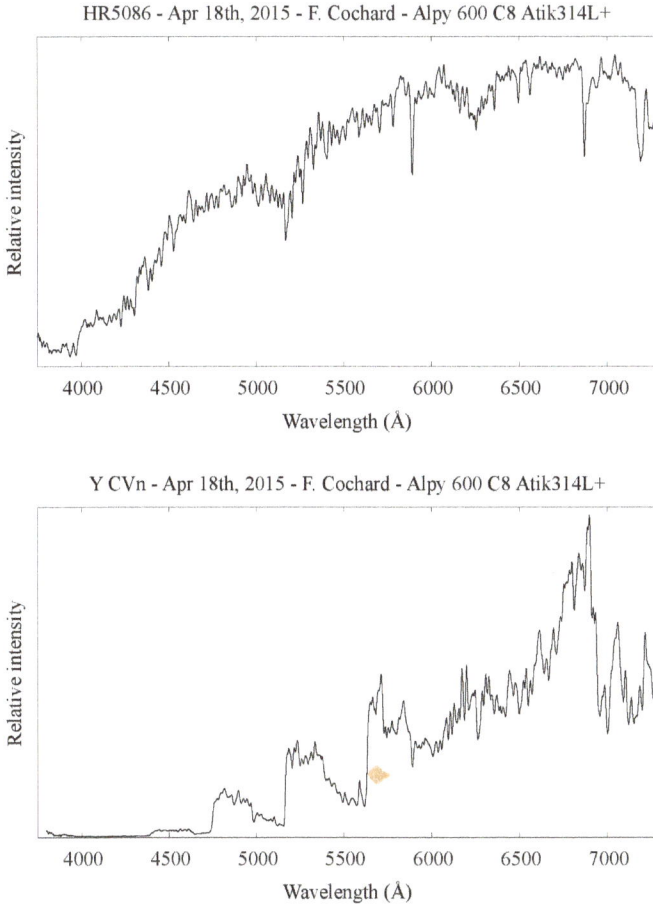

FIG. 1.5 – Continued.

phenomena – and it is a true miracle that this light can travel huge distances (for millions of years at the speed of light!) maintaining this information intact.

The astronomical images (those that we find every day in the magazines) allow us to understand many things about the Universe – motion of the sky and in the sky, magnitude variations of some stars, nebulae, galaxies, etc. But spectroscopy deeply transformed the human comprehension of the sky, giving us a sort of third dimension in the images. A star is not a simple shiny point anymore; it becomes a whole new story to be told. Spectroscopy makes human observations jump from astronomy to astrophysics, that is from contemplation to physical comprehension.

Am I telling you that all of this is accessible to the modest amateur you think you are? Yes, I am.

Nowadays, it takes much more practical common sense than theoretical knowledge to obtain the spectrum of a light source: our mastering of optics has made many obstacles fall, and industrial production allows one to find essential components for affordable cost.

Obviously, an amateur in their backyard will not make the same measurements as a professional observer using a telescope of 10 m diameter situated in an exceptional location – the faintest objects will be inaccessible to us and the precision of our measurements will not always be the same.

But let me tell you that the quantity of stars in the sky – even considering only those accessible to our amateur instruments – is vastly greater than the observing capacity of all of humanity. There are still so many things to discover and understand, and it is a mistake to believe that research is confined in large observatories. Doing amateur spectroscopy means to find again the curiosity proper to the researcher, and it means making totally unedited observations very soon. I could even go on: professional astronomers sure have high performance instruments, but amateurs – in the last few years – have a greater observing capacity. For example, they are the only ones capable of following spectroscopically on a regular basis a sample of a few dozen stars, to measure their time evolution. It is for this reason that the collaborations between amateurs and professionals nowadays are on the rise: it is not a condescending look down from the researchers, but really a collaborative effort.

1.4   Short History

Astronomical spectroscopy as we know it today is the result of a long history in the human knowledge and subsequent discoveries. To cite the beautiful expression from Newton "If I have seen further it is by standing on the shoulders of giants"[4]. I am far from being able to do a historians' job here (many works on this subject exist), but I feel it would still be useful to give some perspective to all the curious and researchers who allowed us to enjoy this marvelous playground today.

As I said before, the beginning of spectroscopy dates back to the work of Isaac Newton (1642-1726), and his experiment of light decomposition through a prism in 1666. He was the one to understand that white light (like the Sun light) can be decomposed in a multitude of pure colors.

The next step came with the Englishman Thomas Young (1773-1829), who understood that light behaves like a wave. He demonstrated this by producing interference fringes using narrow slits a short distance apart. The experiment on interference fringes with Young's slits is now a classic of high school physics lab courses.

The German Joseph von Fraunhofer (1787-1826) invented the diffraction gratings, based on the wave-like behavior of light. The gratings are the heart

---

[4] Letter to Robert Hooke, February 5, 1675.

of most modern spectroscopes, because they allow for a greater dispersion than prisms.

He was also one of the first, together with the Englishman William Wollaston (1766-1828), in 1815, to observe absorption lines in the solar spectrum.

In 1834, the Scot David Brewster (1781-1868) understood that the dark lines in the solar spectrum correspond to absorption from the gas sitting between the Sun and the observer – in the solar atmosphere...and in Earth's.

More or less in the same time, at the beginning of nineteenth century, the Englishman John Herschel (1792-1871) and William Henry Fox Talbot (1800-1877) suggested using spectroscopy for the chemical analysis of substances.

Many people then used this technique in chemistry. For example, the Frenchman Léon Foucault (1819-1868) demonstrated experimentally in 1849 that absorption and emission lines of the same color (same wavelength) come from the same chemical element.

In parallel to these works, in 1842, the Frenchman Edmond Becquerel (1820-1891) was the first to take a picture of the solar spectrum using the invention of Louis Daguerre (1787-1851) dating back to 1839.

The Germans Robert Wilhelm Bunsen (1811-1899) and Gustav Kirchhoff (1824-1887) showed that each chemical element has its own spectrum, and thus that we can use spectroscopy to analyze the chemical composition of any compound. They also established the link – essential for astrophysics – between some absorption lines in the solar spectrum and chemical elements known on Earth.

Kirchhoff went to the point of establishing the three famous laws named after him, which described the three type of spectra – continuous, emission and absorption.

The Sun has been observed many times, because its spectrum is extremely rich, and because its light is abundant – a condition which is crucial for visual, or the then newly born photographic, observations. But we can consider that astrophysics started with the first observation of the spectra of stars others than the Sun – which requires mounting the spectroscope on a telescope or spyglass. Joseph von Fraunhofer was the first to observe, in 1817, the first spectra of bright stars. Between 1860-1880 the observations became more systematic, with the Italian Giovanni Battista Donati (1826-1873), the Frenchman Jules Janssen (1824-1907), the Italian Angelo Secchi (1818-1878), the Englishman William Huggins (1824-1910) and his wife Margaret Lindsay-Huggins (1848-1915). The American Henry Draper (1837-1882) was the first to take pictures of (non-solar) stellar spectra in 1879. It was the epoch of the first stellar classifications (which led, step by step, to the Harvard classification adopted nowadays – the famous O, B, A, F, G, K, M – thanks to Edward Charles Pickering, Williamina Fleming, Annie Jump Cannon and Antonia Maury), and of the discovery of the richness of stellar objects (comets for Donati, planetary nebulae for the Huggins, Wolf-Rayet stars, etc).

In 1848 the discovery by Christian Doppler (Austrian, 1803-1853) and Hippolyte Fizeau (French, 1819-1896) of the effect named after them allowed for the measurement of the motion of a celestial body using the Doppler shift in its spectrum. Among the first to get spectroscopic observations of the Doppler effect, we find again the Huggins' (intrinsic orbit of Sirius in 1868), together with the German Hermann Carl Vogel (1841-1907) who published in 1892 the radial velocity of 51 stars.

These observations lead to a whole period of time in which the aim was the *explanation* of what is seen in a spectrum. We had to wait for all the discoveries regarding quantum mechanics at the beginning of the twentieth century to understand that the origin of spectral lines is in the realm of the infinitely small. In this realm of statistical and quantum physics, I shall mention the work of Max Planck (German, 1858-1947), who determined the spectral distribution of a body as a function of its temperature, Ludwig Boltzmann (Austrian, 1844-1906), Joseph Stefan (Austrian, 1835-1893), Wilhelm Wien (German, 1864-1928), Niels Bohr (Danish, 1885-1962), etc. – the list is long.

As a consequence of understanding the structure of matter, we can now better understand all the information that spectroscopy reveals about the physics of stars. For example, Hans Bethe (German-born American, 1906-2005) established in the 1930s that the energy powering stars comes from the nuclear fusion of hydrogen atoms. into helium, and opened the path to all the discoveries regarding nucleosynthesis (i.e. the production of elements heavier than hydrogen in the interior of stars) – and thus to the history of the Universe. The entire twentieth century is marked by the research on stellar physics and cosmology. These research fields have never been as rich and attractive as today.

Recently, new technologies have revolutionized once again astronomical and spectroscopic observations. More specifically, I am thinking of CCD detectors (and then CMOS), that greatly simplified the set up of the instruments and more importantly, allowed for the transfer of data in the digital realm – helped by growing computing power.

Another innovative technology is worth mentioning: optical fibers. The spectroscopic study of the Sun has been intense since the beginning of the nineteenth century. But we had to wait until the 1960s (which is yesterday!) for access for helioseismology, that is the study of the surface vibrations of the Sun. This requires a very high resolution to detect very weak Doppler shifts. The reward is a much improved understanding of the internal mechanical behavior of our star. Of course, astronomers could not resist the temptation of applying the same technique to other stars, but they came across a technical difficulty: to observe distant stars at high resolution, you need a large telescope. Now, for reasons that will become clearer later, a high resolution spectrograph for a large telescope is necessarily a large instrument – and if mounted directly on the telescope, the mechanical deformations are such that the resolution is strongly downgraded. This obstacle was overcome only in the 1980s with the development of optical fibers. With these, it is possible to physically separate the spectrograph from the telescope, thus removing the problem of mechanical deformations. It is now less than 40 years that we can observe the vibrations of stars – a discipline called *asteroseismology*. Incidentally, the same instruments allowed for the discovery of exoplanets (planets orbiting stars other than the Sun – the first was discovered in 1995 orbiting 51 Peg, by Michel Mayor and Didier Queloz at the l'Observatoire de Haute-Provence).

Nowadays, CCD cameras, high resolution spectroscopy and optical fibers are instruments available to the community of amateur astronomers.

The take home point of this brief historical overview is that this is a young (just slightly more than 2 centuries) and still sparkling adventure. Tackling amateur spectroscopy today is not just being a consumer of well established knowledge, but instead it is tuning yourself into a dynamical community that always wants to improve its understanding of the universe surrounding us.

You can find many books about the history of scientific research. I suggest, for example, the book *The Analysis of Starlight* from Hearnshaw [4].[5]

1.5 Amateur Spectroscopy Today

Since the 1990s, some amateurs (for example in France Alain Klotz, Valérie Desnoux, Christian Buil...) are determined to do spectroscopy with modest instruments. Back then the first step was to build your own instrument: they opened the path. This epoch corresponds to the arrival of CCD cameras and of personal computers. Thanks to these new technological elements, amateurs began to have access to all the ingredients necessary to reproduce (on small scale) the instrumentation of big observatories.

In 2003, Jean-Pierre Rozelot, thanks to some personal connection, organized a CNRS (Centre National de la Recherche Scientifique) school of astrophysics on the island of Oléron, dedicated to collaborations between amateurs and professionals. This almost magic meeting[6] was the starting point of a collective adventure that lead to the creation of the spectroscope Lhires III (and slightly later, to the creation of the company Shelyak Instruments, that still distributes this instrument) and the start of the observational campaign on Be stars with Coralie Neiner's team at the Observatory of Paris – the first true structured Pro-Am collaboration in spectroscopy. To me, this meeting was the first contact with the world of astronomical research.

From this band of passionate pioneers, amateur spectroscopy has been growing little by little, and gathered more and more observers. In many

---

[5] if you can read french, I recommend the great book from Bernard Maitte, *Une histoire de la lumière, de Platon au photon*.

[6] A book has been published after this meeting: *Astronomical Spectrography for Amateurs*, Oléron France, 2003.

European countries, in the United States, and elsewhere, small nuclei of amateurs have appeared, often lead by one or more charismatic leaders.

The development has been particularly fast in France, because amateur astronomy is organized in national associations (AFA, SAF), large local associations (CALA in Lyon, SAN in Nantes, etc.) and some other structures for example Aude-L (and its distribution list[7]) or the forum Webastro[8].

Thanks to these organizations that could spread the information, we could organize or participate in several workshops: Spectro Star Party at the Observatory of Haute Provence (OHP[9]), the Astrophysics School in La Rochelle, "Rencontre du Ciel et de l'Espace" (AFA, in Paris every other year). These meetings are often networking occasions between amateur and professionals. This is how collaborations are born!

At the beginning, most of the professional astronomers looked down on us with a slightly condescending attitude – with of course notable exceptions. But I have always been surprised by their willingness to share, to explain their research – a willingness very much strengthened by having in front of them interested people, and more and more competent too.

Let me give one example among many others. Following the meeting in Oléron in 2003, and the connections that were established there, Jacques Boussuge has worked a lot to develop spectroscopy in the association AstroQueyras (which manages the 60cm telescope in St-Véran). Coralie Neiner (astrophysicist in the Observatory of Paris-Meudon), on her side, worked to get to AstroQueyras a Musicos spectroscope – the same one operating for a long time on the 2m telescope at Pic du Midi (Télescope Bernard Lyot - TBL). Therefore, amateurs found themselves with access to professional equipment – which is very helpful to progress quickly. At that time, we had marvelous help from many researchers from the Observatory of Paris (Coralie Neiner, Paul Felenbok, Claude Catala...).

Spectroscopy as an Educational Instrument

Astronomy has always been a doorway to teach science, and spectroscopy opens a new dimension in that too. One thing is to understand – for example – the motion of a celestial body (planet, star...); but to follow the entire pipeline that goes from image acquisition in the telescope to the determination of a physical quantity is something else. Mechanics, electronics, computer science, physics, chemistry... a large number of disciplines become accessible thanks to astronomical spectroscopy. Whether for your own comprehension, or for didactic purposes, it is an inexhaustible source. To the best of my knowledge, most amateur astronomers doing spectroscopy, do so in this educational spirit, be it a simple hobby, or for didactic purposes.

---

[7] URL: https://fr.groups.yahoo.com/neo/groups/aude-L/info
[8] URL: http://www.webastro.net/
[9] URL: http://www.obs-hp.fr/welcome.shtml

When Amateurs Participate to Research

On top of the educational value of spectroscopy, amateurs can directly contribute to scientific research. Today, the attitude of professional astronomers toward amateur spectroscopy has changed, and even the community of professionals admits that amateurs are capable of quality observations. Scientific publications relying on amateur observations are very numerous. The rapidity in response, availability, and geographical distribution of amateurs allow them to make observations complimentary to those of professionals.

Of course, do not believe that amateurs make the same observations as professionals. In astronomy, the size of the instrument is often directly proportional to its performance, and amateurs will never be able to compete in terms of very high resolution or observations of particularly faint objects. Not all the research lines are suitable for collaborations with amateurs. But it is absolutely crucial to understand that the quality of an astrophysical measurement depends very often more on the rigorous methodology of the observer than on the size of the instrument. A small amateur instrument cannot see object as faint as those accessible to a large professional telescope, but on brighter objects it will be able to produce data of comparable quality. Nowadays, it is common to mix amateur and professionals results – and nobody can distinguish them.

Present day professional research focuses more and more of its resources on bigger and bigger instruments, and this leads to very strong constraints: it is difficult to obtain telescope time (and it has to be planned way in advance), and to justify this time, researchers need to present specific science targets. Amateurs do not have these constraints, and with the only condition of observing relatively bright objects (easily to magnitude 10, and as faint as 15-16 with a little more effort), they can give a greatly appreciated contribution for different kinds of observational campaigns:

- for observations requiring fast responses. For example, nova Del 2013 has been observed very intensively at the time of its discovery (the first spectrum – amateur ! – was realized by Olivier Garde, just two hours after the announcement of the photometric discovery, and 4 hours after the discovery itself);

- for high cadence observational campaign lasting days to a few months. For example, in 2010, the star δ Sco has been the target of intense follow-ups for several months. It is a Be star in a binary, which was then at periastron and showed an unexpected signature of strong interaction between the two stars – observations were not disappointing, and allowed for a very tight follow-up of these unexpected physical phenomena;

- for long term follow-up of large samples of objects. The example of Be stars[10] is emblematic: since 2007, more than 100,000 spectra of these objects have been recorded, and a large fraction of them by amateurs.

---

[10] BeSS : http://basebe.obspm.fr/basebe/

One can of course do spectroscopy without the aim of participating in research programs. But I strongly want you to know that it is possible and inspiring – very often amateur astronomers self-censor themselves by believing that, not being professional researchers, they cannot legitimately contribute to a research work.

Observing Together to go Further

I cannot avoid adding a few lines about what to me is a deep motivation. By approaching spectroscopy, you enter in a new world, which is not yet done amazing you. Whatever the resolution of your instrument is, whatever your case of study, your instrument will allow measurements of an astonishing quality and precision. Even more so since this information is completely inaccessible to our senses – a star remains a simple shiny dot to our eyes.

Your observations will give you a new point of view on stars, and probably this will bring you many new questions.

I want to draw your attention to two important points.

First of all, be aware that, once you understand a few technical aspects described in this book, your measurements will have real scientific value, and they will be useful to the entire community of astronomers. Tell yourself that, often, when you observe spectroscopically, the chances of someone else on Earth observing the same object are very low. Your observation is most likely *unique*. There are too many stars in the sky, and too few observers doing spectroscopy in the World (a few hundreds, at best – a few thousands if we consider also slit-less spectroscopes) to cover the whole sky. You would be amazed at the "little knowledge" of the sky that the community has, if compared to all the observations yet to be done.

The second point is that, in front of this huge observing task facing us, a single isolated person cannot do much. Of course, there are a few prolific observers with an uncommon observing capacity, but they are not more than a handful. And even they have a limited observing capacity in the end. On the other hand, when amateurs observe in a coordinated way, they start a new page in the history of astronomy. Collectively, amateurs can follow a large number of objects, with a temporal cadence inaccessible to any professional observer. At all times, astronomy has been a domain of collaboration between amateurs and professionals. Spectroscopy extends to infinity this domain. It can be seen, for example, through the many programs of collaboration on the rise in the last few years. Some programs, such as the follow-up of Be stars[11] or of cataclysmic and symbiotic stars[12] are long term. Others are

---

[11] BeSS : http://basebe.obspm.fr/basebe/ end ArasBeAm : http://arasbeam.free.fr/
[12] Aras : http://www.astrosurf.com/aras/ and François Teyssier's website : http://www.astronomie-amateur.fr/

time-limited campaigns. Finally some other targets are transient events such as novae or supernovae[13].

This is just the beginning of the history – and we need you to go further. So, take your time to master your instrument and your observations, but get ready to enter in a new adventure which probably goes beyond whatever you dared to imagine.

Everything moves in the Sky

Regularly, when I say that I am doing astronomy, I get the question asking "what's out there for us to see tonight?" – as if you need to wait for the comet of the century to lift your eyes to the sky. In spectroscopy, the "big show" is every night! Lots of objects have a fast evolution – on a timescale between a few hours and a few years. There are foreseeable phenomena, sometimes periodic (for example the behavior of binary stars), and erratic phenomena (Be stars, cataclysmic stars, etc.). Since spectroscopy gives a lot of different physical information (chemical composition, temperature, pressure, velocity, etc.), the variety of spectra accessible to amateurs is virtually infinite. It is even surprising to see very bright stars (Deneb, Rigel, Sheliak, δ Sco, ...) showing strong evolution. In spectroscopy, having images that can be taken twenty years later without any difference is over – your entire life will not be enough to explore everything.

1.6 Starting Spectroscopy

It is time to dive into the guts of the subject. Your mission is to obtain, by yourself, the spectra of astronomical objects.

Book Organization

This book is subdivided in several parts, echoing the "theoretical, instrumental and practical observations". It is likely that some parts will be more familiar to you than others, because of "your personal history in astronomy". Each part is relatively independent from the others, however you will need to be sufficiently comfortable in each for the ensemble to work out fine.

I will start from the theoretical part of spectroscopy: what is light, and what does it tell us about the stars. This is the subject of the first three chapters. Then chapters 4 to 9 are dedicated to the spectroscope and its set up, regardless of the telescope. Finally, the last five chapters deal with observations on the ground.

---

[13] See for example the nova in the Dolphin constellation (August 2013) : http://www.spectro-aras.com/forum/viewtopic.php?f=5&t=682

Do I need to be good in Math?

I choose to put in the text the main equations useful in spectroscopy. These are generally very simple, and they might be useful, for example, to choose the size of the instrument. I suppose there will be some readers with bad memories of their math courses, and to whom the smallest equation is sickening. This is no big deal: just forget the formulae. And for the others, do not hesitate to play with these equations: they will convince you that you are in a very deterministic realm.

Needed Material

It is important to put in practice quickly what I propose here. This means, of course, you need to have the material ready for you to use – be it yours or of your club. The main elements are the following:

- a telescope;

- a mount motorized on both axes, with an automatic pointing function (GOTO);

- a slit spectroscope, with a guiding system included;

- a CCD camera for the spectra acquisition;

- a pointing system;

- a computer to control the ensemble.

Some comments on these materials.

The diameter of your instrument does not matter much: you can do spectroscopy with a refracting telescope of 80 mm diameter. It is not necessary to aim for a large diameter at the beginning, even if, of course, the larger the telescope, the more faint objects you will be able to observe.

The mount, the automatic pointing (*GOTO* function) and the autoguiding capability (i.e. the possibility of guiding the mount from the computer) matter. The autoguiding is not strictly necessary at the beginning, but if you wish to go further than the discovery phase, you will definitely need it. I have seen observers doing spectroscopy with primitive mounts...but it's for people that love challenge and it remains of limited marginal use. This book is not tailored to this or that spectroscope. It can be low or high resolution, the differences will be marginal for the topics I deal with (although they will be mentioned in many chapters). Of course, the set up of the spectroscope itself can differ a lot from model to model, and you will need to use the documentation of your own instrument. But here I describe the general procedure to set up the ensemble of your observing instrument. There is, however, an important caveat: I deal here with slit spectroscopes. Many of the recommendations presented here applies also to slit-less spectroscopes, but the global approach is slightly different. In particular, the Star Analyser[14] case is covered in other books.

I assume you are using a CCD camera for acquisition. The use of a digital camera is also possible (all the advice in these pages applies), but the performance of a CCD camera (more sensitive and cooled) will always be significantly better. In the chapter dedicated to CCD cameras, I will outline the differences between CCDs and digital cameras.

The guiding system is important to make observations lasting more than a few seconds – as it is needed for most stars. In some cases, the guiding can be done using purely visual systems, with the big advantage of a simple set up. But even here, after the discovery stage, a guiding camera will quickly become necessary. Generally, we use a small sensor for the guiding, but you need to check that it is sensitive enough to work in good conditions, especially at low resolution.

You need to be able to point to faint objects with your instrument. I will spend some time on this issue in chapter 11, but I want you to be already aware that it can be very handy to have a third camera, entirely dedicated to pointing your telescope to the required field.

Software Tools and Problems

Practising spectroscopy requires intensive use of computers. Cameras and telescope control, use of digital sky maps, self-guiding, data reduction and managing, web searches: all of these require a computer. A good handle on the software tools is therefore necessary to avoid turning spectroscopy into a nightmare. It is frequent to see observers curse at the complexity of spectroscopy, while they are just facing driver issues, or image acquisition software not adapted to their needs.

I can summarize here the main difficulties you might face.

- First of all, you need to be comfortable with the basic computer operations: file and folder manipulations, internet connection, etc.

- Software installation and configuration. There are many pieces of software to be installed: image acquisition (for the spectroscope), guiding and self-guiding, data reduction, sky maps, etc. Often this requires root privileges (i.e. "Install as Administrator" privileges) on your PC.

- Driver installation, for the cameras, and for the telescope. It is often a complex matter, because the installation process is specific to each device manufacturer.

- Installation and configuration of tools such as a USB-serial converter. In reality, most of the tools we use often come with a serial connector (or RS-232), but these connectors do not exist anymore on modern PCs.

---

[14] Shelyak Instruments : http://www.shelyak.com/rubrique.php?id_rubrique=4

On the other hand, PCs all have multiple USB ports, and we can thus employ converters from one format to the other. Such converters are easy to find on the market. However, they are often fragile components, with a limited life-span: you need to be able to diagnose problems with them.

Even with experience, you can waste a lot of time on this or that computer issue. You need to keep in mind that the complexity arises because of the large number of pieces of equipment that need to cooperate.

The more you try to automate your installation, the more these computer issues will become dominant. For this reason, it is important to start with simple tools, independent from each other.

Required Computing Power?

With all this computing activities, you might think that you need a super powerful computer. This is not the case: today, the most common laptop can control many cameras and a telescope. Using a more powerful machine will always be beneficial (faster reactions), but it is not necessary. It might be useful to separate the different functions mentioned above, and to use more than one computer. Not because of a power issue, but simply to divide the complexity of the installation. In my present personal configuration, I use my main PC for spectra acquisition and data reduction, a second small PC (netPC, bought for a few hundred euros) for the autoguiding (it is therefore connected to the guiding camera and the telescope), and a screen to see the image of the pointing field (see below). On top of the advantage of separating each function, I also get a larger screen surface that allow me to avoid switching between the different software interfaces (this could also be achieved with an external monitor). Thus, for example, I always have the guiding image visible, and I can check that the telescope is perfectly on target without effort.

What Instrument to what Measurement?

I want to remind you here about an obvious thing: spectroscopy is not a science in itself, but rather a tool to access the science of stars in the sky. A tool is meaningful only in perspective of its use – the hammer was not invented to look good, but to hammer nails. You will see that spectroscopy can "hammer many nails", because of the large information content of the light of a star. But unfortunately, there is no ideal spectroscope, capable of covering efficiently all the astrophysical domains. You will have to make choices and compromises. During the whole process, ask yourself regularly these essential questions. "What am I looking for?" "Is the instrument I am using adapted to my research?" "Why am I doing what I am doing?" . Tell yourself that the advice I give here can vary significantly depending on your objective.

Practice Regularly

I would love to observe more often, more regularly...and I know I am not the only one. Some observers have an observatory and very favorable conditions for a regular practice of astronomy. But they are not the majority of people – who are much often constrained to observe only a few times per year, at most.

I need to be very clear: to do spectroscopy once or twice a year is an almost impossible mission, especially while you are still setting up your instrument. Mastering your instrument requires regular interventions, to dominate little by little each single element. If you don't practice often enough, you will forget what you learned last time, and you will never see the end.

Think about your practice on the long term.

And take advantage of the fact that spectroscopy can be practised in town (I know of assiduous observers who often use their instruments in a urban environment). Actually, it turns out that with a slit spectroscope, you record the spectrum of the background sky – and thus of the possible light pollution – simultaneously to the stellar spectrum; it is therefore possible to post-process the spectra removing light pollution to a certain extent.

I live in an apartment, in an urban environment – conditions in which one would legitimately think it is impossible to observe. Nevertheless, I have installed my telescope on my balcony... and this allowed me to make huge progress! I only see an infinitesimal part of the sky, and even that is strongly polluted, but I see enough stars to work, little by little on the different points treated in the following pages. Thus, when I have the occasion to observe in a better environment, I am ready!

A Time for each Thing

It is an element that became evident to me, after discovering it little by little from the contact with professional astronomers. In observational astronomy, there is a time for each thing. There is a time to set up and tune your instrument. Then, there is a time to observe. These two times are incompatible, mutually exclusive. Amateurs often have the tendency to mix them up, and it is a certain source of complexity. Moreover, the majority of the set up on the instrument should be done during the day. The main exception is polar alignment, which requires seeing some stars. Unfortunately, it is common to see amateur observers waiting for the night to install their instruments...it's a waste of time! Setting up at night means losing precious observing time.

I invite you to make the effort of separating these two times. If you do the start up and set up operations, or you modify your instrument, do not consider your measurements reliable. Conversely, when you observe, forget your instrument – what matters are the measurements you are taking.

Of course, there will always be exceptions. But if, when you set up, you tell yourself that your objective is to have an instrument that will allow for easy and reproducible observations, you will gain a lot of time. During your observations, avoid all modifications to your instrument – the quality of your measurements will improve.

Navigating between two Pitfalls

During your learning of spectroscopy, I invite you to check on two common pitfalls.

Do not try to put all the elements – telescope, spectroscope, guidance, cameras... – together at once to obtain a spectrum more quickly. The probability for everything to go smoothly the first time is infinitesimal... and even if this happens, you will not learn much from it. Take the time to master each single element.

Conversely, don't be a perfectionist in all settings. The aim is to obtain spectra, not to drown in the instrument settings. It is frequent to see observers getting lost in settings way beyond what is needed. It is only when you obtain the first spectra that you can understand the relative influence of each parameter.

Therefore, there is an balance to find to get quickly to the result – a valid spectrum – without confusing being fast with being in a hurry. I am a fervent believer of continuous improvement: you need to quickly get your first spectrum – which requires to set up the main technical elements – so that later you can improve, little by little the result, taking care of measuring objectively the quality obtained to see progress.

Expose yourself to Critics

After the first spectrum, consider comparing it to others and share your results[15] to encourage other beginners. The more we are, the better and faster we move on. Do not fear criticism, go and look for it!

---

[15] See for example the Aras forum: http://www.spectro-aras.com/forum/index.php

Chapter 2

Light

In this chapter, we are going to look at the nature of light and how it is produced and modified by various physical phenomena.

Light can be represented in two complementary ways : it is a wave and it is a beam of particles.

2.1 Light is a Wave

First of all light is an electromagnetic wave, that propagates (in a vacuum) along straight lines and at constant speed. This wave is represented by a sinusoidal curve (fig. 2.1), and it is characterized by its wavelength λ (pronounced "lambda"), which represents the distance between two crests of the wave.

Since this wave propagates at the speed of light c (approximately 300,000 km s^{-1}), we can also characterize it using the frequency of its oscillations ν (pronounced "nu"), which is measured in Hertz (Hz, vibrations per second). The relationship between frequency and wavelength is $\lambda = c/\nu$ or, equivalently, $\nu = c/\lambda$.

This electromagnetic wave transfers energy and the higher the frequency of the electromagnetic wave, the more energy it carries.

There are no bounds to the wavelengths possible in free space. The most energetic rays encountered in astrophysics are the γ rays (pronounced

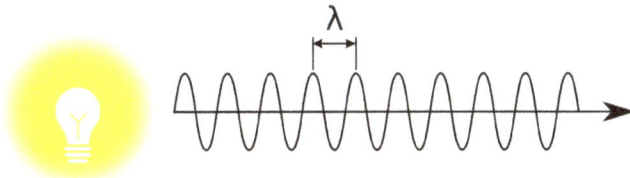

FIG. 2.1 – Representation of an electromagnetic wave.

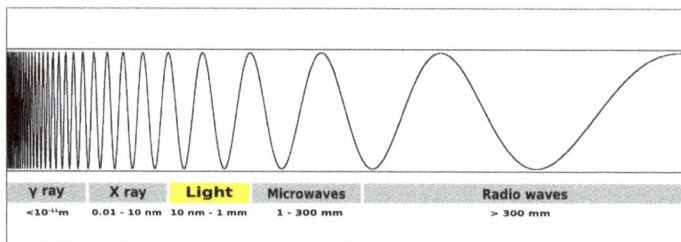

FIG. 2.2 – Electromagnetic Spectrum.

"gamma"), whose wavelength is less than $10\,\text{pm}$ (picometers – $1\,\text{pm} = 10^{-12}\,\text{m}$). On the opposite side, the least energetic rays are radio waves, with a wavelength exceeding $300\,\text{mm}$ (up to several tens to thousands of meters). Therefore, the ratio between the longest and shortest wavelength is higher than 10^{15} (fig. 2.2).

> The *official* unit of measure for wavelength in the international system is the nm (nanometer : $1\,\text{nm} = 10^{-9}\,\text{m}$). However, in your practice of spectroscopy, you will find both nm and Angströms (Å). One Angström is equal to $10^{-10}\,\text{m}$, therefore the relationship between the two is easy: $1\,\text{nm} = 10\,\text{Å}$. Regularly, purist fight over this, but actually this duality is not really a problem – you get used to it quickly. In this book, I chose to use nm – but in some figures the wavelength is given in Å.

Right in the middle of this range, we find what we call "the light" in everyday experience, between $10\,\text{nm}$ and $1\,\text{mm}$ (i.e. $10^6\,\text{nm}$). This much narrower range of the electromagnetic spectrum is in turn subdivided in three parts (fig. 2.3) :

– the ultraviolet, from $10\,\text{nm}$ to $380\,\text{nm}$[16];
– the visible, from $380\,\text{nm}$ to $750\,\text{nm}$ (often rounded to 400-700 nm);
– the infrared, from $750\,\text{nm}$ to $10^6\,\text{nm}$.

Light in the visible range is no "different" than any other wavelength range, and it is "special" only because the human eye is sensitive to this extremely narrow window. The infrared domain is very extended, and is separated into near infrared (NIR, until roughly 3,000 nm), mid and far infrared, depending on the type of detector.

For what concerns us here – amateur spectroscopy – we will mainly deal with the visible and near infrared domain: these are the ones in which present

---

[16] in Astronomy, UV starts from 300 nm, which is the lower end of the atmospheric window

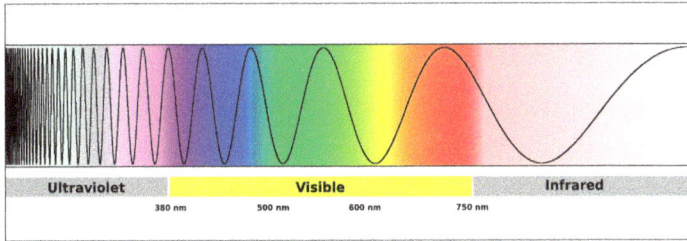

FIG. 2.3 – Electromagnetic spectrum of light.

detectors (CCD and CMOS) are sensitive. Thus, we generally work on wavelengths between 380 nm and 1,000 nm, keeping in mind that beyond 750 nm, our electronic cameras are sensitive, but not our eye.

In the visible domain the wavelength corresponds to the color. This means that not only are our eyes sensitive to electromagnetic radiation in this wavelength range, but that they can also distinguish with an incredible precision the different wavelengths in this range.

The shortest visible wavelengths correspond to the color deep violet – and they are also the most energetic. The longest correspond to red. In between these two extremes, the eye can distinguish many colors: purple, blue, green, yellow, orange, red – in reality, a continuum of colors.

The light we see everyday – from the Sun, or from artificial illumination – is a bit more complex than this. It is not a unique color, but of a mixture of a multitude of colors. Isaac Newton (him again!) demonstrated with a very simple experiment that white light could be decomposed into all the colors in the rainbow.

Why is it that the human eye is particularly sensitive to the range 400-700 nm ? Is this range somehow special? Yes: it corresponds to the domain in which the Sun emits most of its energy (we'll come back to this later). The sensitivity of the eye to this domain is a consequence of *evolution.*

Waves and Interference

When waves combine, they produce interference phenomena. Suppose we have two identical waves, with the same frequency. If these are perfectly in phase, they add (fig. 2.4).

On the other hand, if they have opposite phases, that is one is half of a wavelength late with respect to the other, then they will cancel each other (fig. 2.5).

All the intermediate stages (phase difference of any other value) will have intermediate effects, between addition and mutual cancellation. This is the mechanism that leads to interference, for example in the experiment of Young's slits: imagine two narrow slits very close to each other, through which two light waves pass and shine on a screen (fig. 2.6).

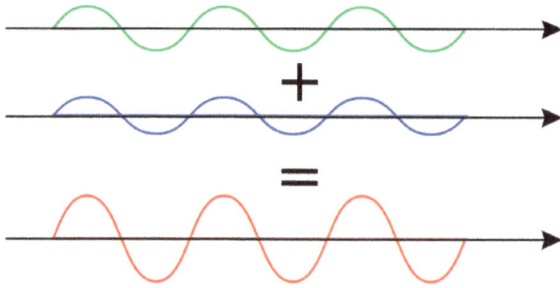

FIG. 2.4 – Sum of two waves in phase.

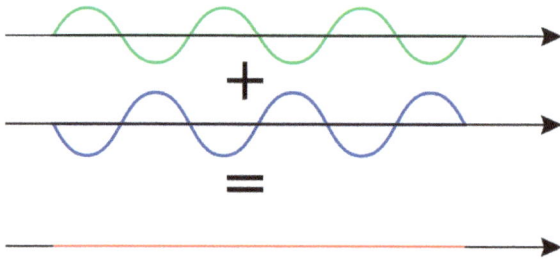

FIG. 2.5 – Sum of two waves in phase opposition.

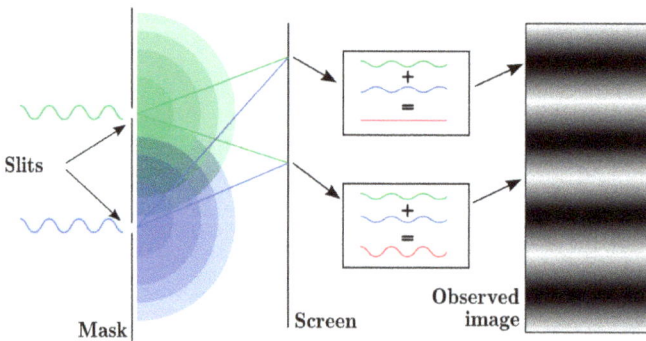

FIG. 2.6 – Interference phenomenon

For a given position of the screen, each ray takes a path of a specific length – easy to work out with trigonometry. The difference between the optical paths will be an integer multiple of the wavelength for some points on the screen, in which case, the waves of the two rays have the same phase and add to each other, so the light on the screen is intense. For other points on the screen, there will be a mutual cancellation of the two waves, and these points are dark. Thus, when we look at the whole screen we see a series of dark and bright fringes.

2.2 Light is a Particle

Now that we have looked at light as a wave, let us look at its other representation: light is also a beam of particles – photons. The photon is an impulse, like a "particle of light". It has no mass, but transports energy and momentum.

It seems to be very different compared to the wave but nevertheless, the two representations are complementary and both have been formally demonstrated.

To each photon, we can associate an oscillation of the same frequency ν of the corresponding electromagnetic wave. The relationship between the photon energy E and the frequency of its oscillation is given by the relation $E = h\nu$ where h is the Planck constant (approximately equal to 6.62×10^{-34} J s), and for this reason we talk about the energy.

Matter is made of small particles, atoms, which are in turn made of smaller components: protons, neutrons, electrons, etc. Each atom – hydrogen, helium, carbon, silicon, etc. – has a very specific number of electrons, the same as the number of protons in the nuclei.

The more massive the atom, the more electrons surround the nucleus. The electrons are not randomly placed around the nucleus: they have particular orbits, each of which can accept a precise number of electrons. For example, the first orbital can host two electrons, the second eight, etc. When an electron is in an orbital, it has a specific energy level.

There is another important element to understand; which requires a short detour through quantum mechanics. This one tells us that an electron can, under certain circumstances, change energy level, and it needs to exchange the energy it gains or loses with the external environment...in the form of a photon. An extraordinary thing – with direct consequences for spectroscopy – is that this energy transfer is not continuous, it is sudden and without intermediate stages. The energy given or absorbed by the electron corresponds exactly to the difference in energy of the initial and final orbital of the electron.

This change in energy levels in the atom produces light with $\Delta E = h \cdot \nu$, that is with a wavelength given by the relation $\lambda = c/\nu$, so $\lambda = c \cdot h/\Delta E$.

Since each atom of a specific element has its own structure, all the possible combinations of energy changes for the electrons are specific to that element. This also determines a specific wavelength and frequency for each photon, either absorbed or emitted, that is specific to each energy transition. We touch here a truly fascinating aspect of astrophysics: we are going to work on light that has traveled for *astronomical* distance and time, but this light still carries information about the atoms in the source, typically a star. This relationship between the infinitely large and infinitely small is incredible: when we observe a star, we access precise information about the atoms that compose it.

Neither of the two representations – wave and particle – is "more correct" than the other, both should be regarded as two simplifications of a more complex reality.

When we do spectroscopy, we need both representations; neither of them would explain completely the observed phenomena. The wave model is very effective to explain optics, while the photon model and the related atomic model explain the origin of absorption and emission spectral lines that we find in spectra.

2.3   Making Light

After describing the nature of light, it is time to ask ourselves how to produce it. There are mainly two ways – once again they will coexist. The first way is to heat up matter, any matter. The second way is to excite atoms.

Heating a Body

The heating translates into an emission of light at all wavelengths, with an energy distribution (as a function of the wavelength) which only depends on the temperature. Max Planck established the law (named after him) describing the distribution of the energy emitted as a function of the wavelength, for any given temperature. The formula can be impressive, but it is easy to calculate (for example in a spreadsheet, to make a plot):

$$L_\lambda = \frac{2hc^2}{\lambda^5} \frac{1}{e^{\left(\frac{hc}{k\lambda T}\right)} - 1}$$

where:

- L_λ : Spectral radiance (in W m^{-3} sterad^{-1});

- h : Planck constant (6.62×10^{-34} J s);

- c : light speed (roughly 3×10^8 m s^{-1});

- λ : wavelength (in m);

- k : Boltzmann constant (1.381×10^{-23} J K^{-1});

- T : body temperature, in Kelvin.

It is important to understand that this phenomenon does not depend on the composition of the body: be it gas, wood, stone, iron, air, water – and even biological matter – the law remains the same.

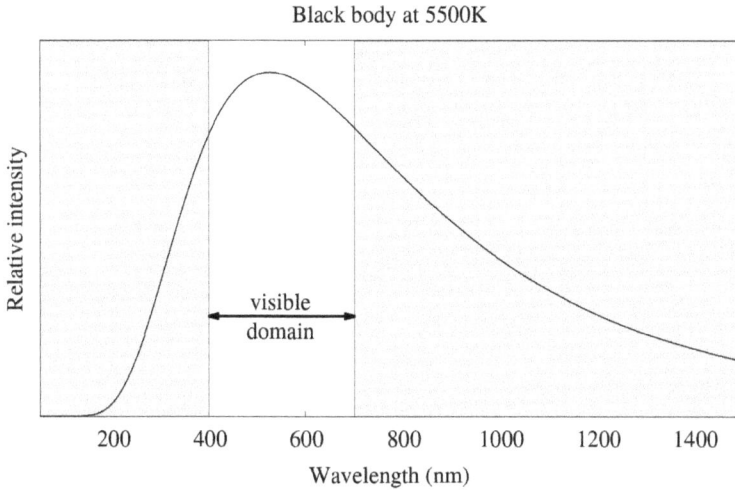

Black body at 5500K

FIG. 2.7 – Planck Profile, or Black Body profile.

When we plot this law, we obtain a characteristic "skewed" shape. For example, figure 2.7 shows the Planck profile for a body at temperature 5,500 K, corresponding roughly to the temperature of the surface of the Sun.

I suggest you familiarize yourself with this law, and plot it with different temperatures. There is always a maximum intensity, whose wavelength varies with the temperature. The following relationship (the Wien law) gives the wavelength of this maximum:

$$\lambda_{max} = \frac{2.898 \times 10^{-3}}{T} \text{ (with } \lambda_{max} \text{ in m, and } T \text{ in K)}$$

In the same way, we can determine the temperature of a body with known wavelength of maximum emission:

$$T = \frac{2.898 \times 10^{-3}}{\lambda_{max}}$$

We have seen that the visible spectral domain goes from 380 nm to 750 nm. We can calculate the temperature of sources with a maximum at each of these wavelengths: this tells us the temperature range for which the maximum of Planck's curve lies in the visible domain.

With $\lambda_{max} = 380$ nm (i.e. 380×10^{-9} m), we obtain:

$$T = \frac{2.898 \times 10^{-3}}{380 \times 10^{-9}} = 7,626 \text{ K.}$$

Similarly, with $\lambda_{max} = 750$ nm (i.e. 750×10^{-9} m), we get:

$$T = \frac{2.898 \times 10^{-3}}{750 \times 10^{-9}} = 3,864\,\text{K}.$$

Sources with a maximum of the emitted intensity in the visible domain must therefore have a temperature between about 3,800 K and 7,600 K. Such sources are not common in everyday life – which also explains why at our scale objects don't seem to be luminous (they are in the infrared). There are however some notable examples: a flame, the filament of an incandescent light bulb, for example – these reach a few thousands degrees (the filament of an incandescent light bulb can reach 2,800 K, for instance). Without spoiling much of the next chapters, be aware that the temperatures of stars cover this range – and this is why we see stars shining.

Keep in mind that the light produced by heating a body will always have a continuous spectrum, with the relative intensity of colors depending only on the temperature.

The spectral profile from Planck's law is commonly called a "Planck profile" or "Blackbody profile". It is a bit surprising to say Black body for a body shining light, but there is a simple explanation. We have seen that at ordinary temperatures on the human scale, bodies do not shine light on their own. However, we do see the objects that surround us. This is because they are illuminated by an external source, generally the Sun or an artificial light and what we observe is the reflection of this light. The Planck profile corresponds to the emission proper to a body, not to the reflection. To observe a pure Planck profile we need to screen any external light source – to dive into the absolute black. This is where the "Black" in the name "Blackbody profile" comes from.

Exciting Atoms

The second way of emitting light consists of exciting atoms. We have seen that in a simplified representation of the atomic structure, the atoms are made of a nucleus and a cohort of electrons orbiting it. Exciting an atom consists in stripping off some electrons, or changing the energy level of one or more of them. This operation requires energy, and the energy level of the atom increases. Once an atom is excited, it naturally tends toward its equilibrium state, therefore, it will substitute or re-position the electrons, emitting the excess of energy as a photon with a frequency characteristic of the corresponding electronic transition.

In this representation, I have described an electron freed from its atom, but this phenomenon never occurs in isolation: we need a statistical approach. In reality, the number of transitions that can happen in a small piece of matter is huge, and each time an electron is freed from its atom, it will soon encounter another on its way.

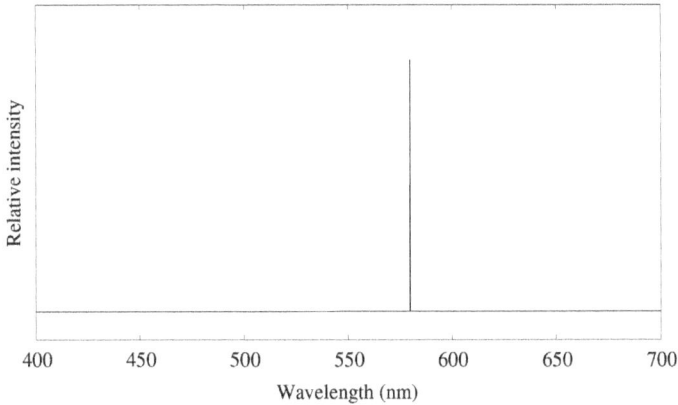

FIG. 2.8 – Theoretical representation of an emission line corresponding to an electronic transition.

In practice, the more probable a transition is, the more it will happen. Since an atom can have a lot of electrons, many transitions are possible, sometimes a very large number. All these transition will happen, proportionally to their probability.

The ensemble of the electronic transitions from a given energy level combine to give light emissions at specific frequencies. This translates into an emission of light at a precise wavelength, and an emission line in the resulting spectrum (fig. 2.8). The bulk of the spectrum is dark, but a bright line is visible.

In practice, as we saw, it is never just one transition that occurs, but the ensemble of all the possible transitions for the atom under the specific physical conditions, and the spectrum is really just the ensemble of the lines (fig. 2.9). The number of lines increases with the complexity of the atom, since the more complex the atom is, the larger the number of possible transitions. The hydrogen atom – the simplest there is – has only a few lines in the visible, while the atom of iron, for example, teems with lines (you will see both in stellar spectra).

I said that we can make light by exciting atoms. Very well, but...how do we excite atoms? It's not easy to grab it and shake it – an atom is too small for that. Nevertheless, the idea of shaking is right: we have to give it just enough energy to strip one electron or change its energy level[17]. We have seen that such energy transition cannot happen – because of the laws of quantum mechanics – unless the energy provided corresponds to the energy difference of the electronic transition.

---

[17] If this is done by means of photons, the energy lost by the light beam will be absorbed by the atom. This energy is then re-emitted by the de-excitation of the atom.

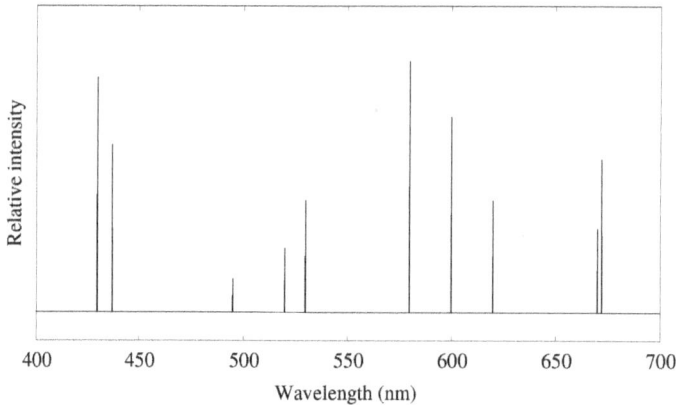

FIG. 2.9 – Theoretical representation of an atom emission spectrum.

Let me try an analogy with everyday life. You certainly had the occasion to shake a tree to make its fruits fall. Imagine a tree that allows the fruits to fall only at the very specific "shake" frequency it needs. A little bit higher or lower, nothing happens. But if you reach exactly the right frequency, then all fruits fall at once. In practice, when you shake the tree, you give it a certain frequency, but this will be transmitted to the fruits with variations depending on the mechanics (resonance) of each branch. If you are far from the right frequency, nothing will happen, but the more you approach it, the more fruits will fall, because local effects can modify slightly the frequency at the fruit. At the nominal frequency, more fruits will fall, but in reality there is a range of frequencies around it for which you will have the desired outcome.

To get back to our atoms, we can "shake" matter, for example using an electric field to make it vibrate – as an earthquake shakes all buildings to a region. Choosing the right frequency for this field (which is easy nowadays), we can excite, for example, a gas: this is exactly what happens in a energy-saving lamp.

Therefore, the general mechanism is the following: we input energy in a material, which excites some of its atoms (i.e. moves their electrons from their rest state). Then, these electrons naturally go back to their rest state, emitting light at a very specific wavelength, characteristic of the corresponding electronic transition.

By heating matter, we obtain a continuous spectrum. By exciting atoms, we obtain a discrete spectrum, made of isolated lines. Two ways of making light, two very different physical phenomena...that we will be able to recognize and distinguish in the observed spectrum! If you see a continuous spectrum, you know it was made by heating, and you can measure the source temperature. If you see an emission lines spectrum, you know it was produced by atomic excitations, and you can recognize the chemical element(s) present in the source.

Combining the Two

We saw that we can produce a continuum spectrum heating any material, and that we can also produce light exciting atoms – for example with an electric field. But light itself is an electromagnetic field, thus if we shine light on a material we can create excitation! Moreover, if we use a source of continuous light (for example with Planck's spectrum), we will have all the possible frequencies, and among them there will be some allowing to remove electrons from the atoms.

Let's imagine a very simple device: a continuous light source (for example an incandescent light bulb) shines on a transparent box containing gas. Among all the wavelengths of the source, some will excite the atoms. This energy will thus be absorbed by the gas, and the photons entering the box with these wavelengths will not exit on the other side. If we look at the spectrum after the gas-containing box, we will find the continuous spectrum, but some lines will be missing – we call these absorption lines. Of course the lines that are missing are exactly the same the gas can emit, since they correspond to the same electronic transitions, but in the opposite direction.

The gas particles that absorbed some energy, will quickly give it back, by letting the electrons fall into their rest state again. But now, the emission of light is distributed in all directions. This is called scattering.

Thus, we are in the following configuration: a light beam is projected on a gas and essentially crosses it without being modified, except for a small portion which is absorbed and then re-emitted in all directions. Consequently, if we look along the axis of the beam through the gas, we see the initial continuous spectrum with some lines missing (absorption lines), while if we look at the gas in the direction perpendicular to the beam, we see an emission line spectrum. In total, if we measure the energy transmitted all around the gas, it is correctly equal to the energy input, but not all the energy is focused in the beam crossing the gas.

These different phenomena thus lead to different kind of spectra, and we can observe in nature continuous spectra, absorption line spectra and emission line spectra. And all the possible combinations. You will rapidly see that the objects in the sky offer an amazing variety of spectra – and you can begin to feel how the observation of these spectra allows to make physical measurements on their sources: this is the power of astrophysics.

2.4 Shifting a Spectrum

For the sake of having a complete physical description of the light, I still have to mention a physical phenomenon that can *modify* the light after it

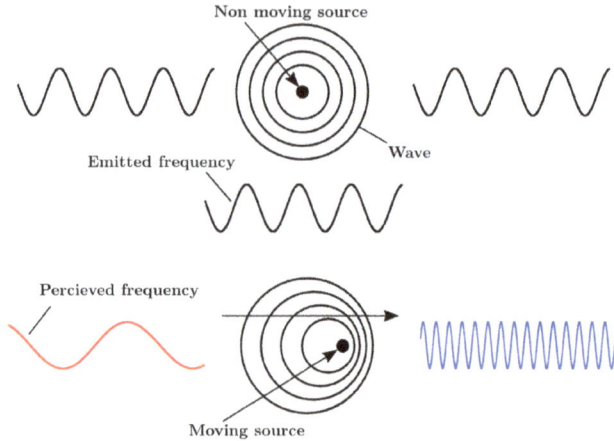

FIG. 2.10 – The Doppler effect.

has been made. It is the Doppler-Fizeau effect (commonly called just Doppler effect) (fig. 2.10). This effect is often presented relying on a common experience anybody has had: when you hear an ambulance on the street, the perceived frequency changes depending on whether the vehicle is approaching or moving away. When the vehicle is approaching, the siren is more high pitched, and becomes lower and lower as the vehicle moves away. If the ambulance stops right next to you, the frequency is intermediate. This example uses a sound instead of light, but the analogy holds completely – it is the same phenomenon. On the one hand there is the frequency emitted by the source, and on the other the one perceived by the detector. If the two are not moving with respect to each other – no matter how distant they are – the two frequencies are identical. But, if the detector is approaching or moving away from the source, the perceived frequency is modified by the displacement velocity.

I will not spend time on the physical explanation of this phenomenon. It comes from simple geometrical considerations – you will easily find a lot of literature on this subject. However, I think it is interesting to point out a few orders of magnitude for the applications concerning us.

It is necessary to underline that when we mention the displacement velocity between the source and the detector, we are only interested in the *radial* component of this velocity, that is the component of the velocity along the direction connecting the source and the detector. If the detector goes around the source, staying at constant distance, we talk about tangential velocity, and in this case, no Doppler effect is detected. A simplified way of looking at this, is to talk about approaching or departing velocity – when we turn around the source, we do not approach it or depart from it.

The mathematical expression for the Doppler effect is below. If λ_{em} is the wavelength of the emitted light, then the wavelength of the observed light, λ_{obs} is

$$\frac{\lambda_{obs}}{\lambda_{em}} = 1 + \frac{v}{c} \quad \text{or} \quad \frac{\Delta\lambda}{\lambda} = \frac{v}{c}$$

where c is the speed of light, and v the velocity of the detector with respect to the source (in the same units of m s^{-1}). We see from this formula that, the higher the displacement velocity, the more important the effect is. In particular, if the displacement velocity is zero, we correctly find:

$$\frac{\lambda_{obs}}{\lambda_{em}} = 1 \quad \text{so} \quad \lambda_{obs} = \lambda_{em}$$

Imagine I am driving my car toward a red traffic light at 50 km hour^{-1} (i.e. 13.8 m s^{-1}). Since I am moving toward the traffic light, the velocity is negative (it would be positive if I were driving away). Since the traffic light is red, we can consider its light to be emitted at a wavelength of approximately 650 nm. The wavelength I perceive for this same light is

$$\lambda_{obs} = \lambda_{em} \times \left(1 + \frac{v}{c}\right) = 650 \times 10^{-9} \times \left(1 - \frac{13.8}{3 \times 10^8}\right) = 649.99997 \times 10^{-9},$$

that is 649.99997 nm.

Our calculations changes only the fifth decimal place of the wavelength. We might as well say that for any velocity "humanly possible", the shift will be totally imperceivable to the naked eye. More in general, the Doppler effect on light is not perceived on the human scale, since the velocities of our everyday life are extremely smaller than the speed of light.

But let's do a small exercise of science-fiction. Suppose now I drive at 168,480,000 kmh^{-1}, that is 46.8 × 10^6 m s^{-1}. The same calculation as above, with the new values yields an observed wavelength of

$$\lambda_{obs} = 650 \times 10^{-9} \times \left(1 - \frac{46.8 \times 10^6}{3 \times 10^8}\right) = 548.6 \times 10^{-9}, \text{ that is } 548.6 \text{ nm.}$$

Now, the initial color is vastly altered, since starting from a red light (at 650 nm), I see a green light at 550 nm. If you do the experience for real, you'll need to convince the police that the light was really green when you crossed the intersection...

Therefore, the entire observed spectrum can be shifted significantly. This creates a singular problem: when I measure a line at a certain wavelength, how can I know its real wavelength? If we do have only one line in the spectrum and no other information, we can't work anything out from it, and astrophysics will have to stop here. But in practice, we always observe many lines, and it is the pattern from the ensemble that allows us to identify the lines – because this pattern is not modified by the Doppler shift. What would be in a certain way a real practical obstacle becomes then an extraordinary opportunity: not only can we find the chemical elements in the spectrum, but once their lines have been identified, we can infer the displacement velocity of the source along the line of sight!

We will later see that while this effect is imperceptible on the human scale, a whole branch of astrohysics is devoted to this.

Doppler Effect: Order of Magnitude

I told you that the Doppler effect on light is generally not detectable on the human scale. I feel it is useful to keep in mind an order of magnitude in the following. Let's consider a shift of 1 Å, that is 0.1 nm. If we go back to the formula for the Doppler shift, and transform it to express the displacement velocity as a function of the wavelengths, we get:

$$v = c \times \frac{\lambda_{obs} - \lambda_{em}}{\lambda_{em}}$$

For a sample wavelength of 550 nm, a shift of 0.1 nm corresponds to a velocity of:

$$v = 3 \times 10^8 \times \frac{0.1}{550} = 54{,}545 \text{ m s}^{-1} \approx 55 \text{ km s}^{-1}$$

This is a good scaling to keep in mind: a displacement velocity of 50 km s^{-1} translates to a Doppler shift of about 1 Å (0.1 nm) around wavelengths of 550 nm.

2.5   What the Human Eye Sees

At this stage, it is useful to look at how the human eye detects the phenomena described above. The retina of the eye is a light detector, which we saw is capable of distinguishing colors with a great precision. Nevertheless, it cannot detect everything, and this is why we need spectroscopes.

The human eye can be considered as the ensemble of three separate cell types (our detectors), the cones (fig. 2.11). Each of the three cones detectors is sensitive, roughly, to one third of the visible spectrum – blue, green and red. Since the visible spectrum spans roughly 300 nm, each detector covers roughly 100 nm (centered in 440 nm, 530 nm, and 600 nm for the blue, green, and red, respectively).

For each of these domains, the eye can make a precise intensity measurement, relative to the intensity in the other domains. It is this kind of sensitivity that allows to our eyes to detect hundreds, possibly thousands of colors.

Nevertheless, the eye is not able to distinguish variations *within* these domains. Imagine a line shifted from 510 nm to 520 nm: to our eye nothing happened. Similarly, if a body emits a couple of dozens of lines between 600 nm

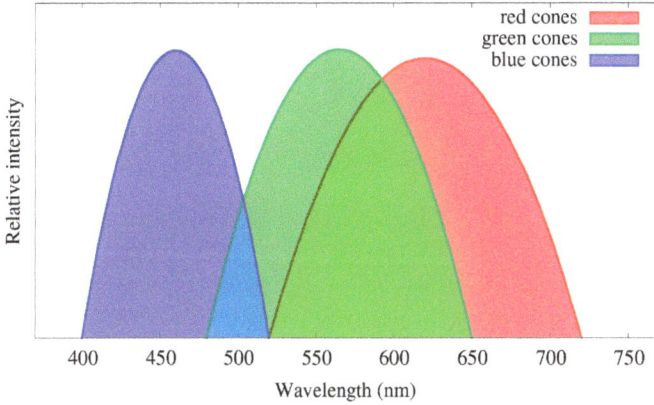

FIG. 2.11 – Spectral domain covered by each type of cones of the human eye.

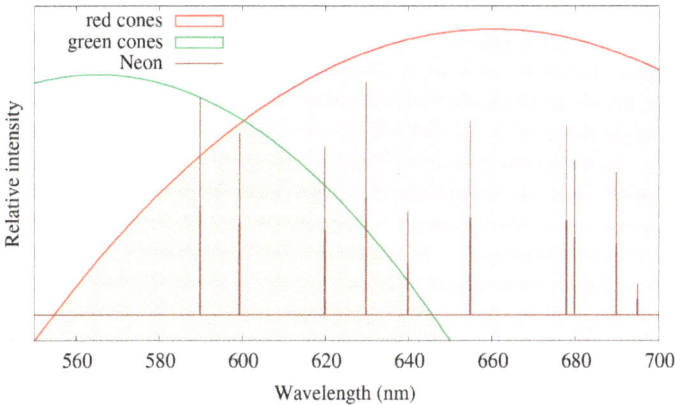

FIG. 2.12 – The human eye cannot distinguish emission lines

and 680 nm (as neon does, for example), our eye will be incapable of distinguishing them (fig. 2.12). In practice, a neon lamp looks red to us, without any perception of the richness of the underlying spectrum.

Thus, without instruments, we are capable of determining with great precision the energy distribution of a light source in the three bands red-green-blue, but we cannot go beyond. But we have seen that the spectra of natural light sources can be much richer than this – and this richness will be accessible only if we apply to our eye a "prosthesis": the spectroscope.

Let me draw your attention to this point: the spectroscope allows us access to information which is otherwise inaccessible to our eye. It is not only a matter of scale, but really of design. When we use a telescope, we amplify the light received using a collecting area much larger than our eye. Were our eyes bigger (and could they collect information for longer), we would not need telescopes. It is not the same in spectroscopy: by construction, our eye cannot detect thin spectral lines.

2.6 Atoms and Molecules

Before closing this chapter, I want to add an element that has often caught my curiosity. I have talked about a simple atomic structure, with a central nucleus and electrons spread around it. Atoms can be excited exchanging energy in the form of photons with the external environment. But in practice, in many cases atoms are not isolated. They are often grouped in molecules; it is through electrons that molecules hold atoms together (in covalent bonds). Molecules are assemblies of atoms, and these assemblies have their own mechanical behavior: they have modes and proper resonance frequencies, and they can also exchange energy with the environment.

The energy levels in molecules are typically lower than in atoms, but they are still quantized, and produce emission and absorption lines. In practice, while many things happen in the visible range for atomic spectra, it is mostly in the infrared that information about molecules is hidden (with some notable exceptions – for example we find molecular lines in the spectra of cool stars).

Molecules are a bound assembly of atoms, and like a mechanical vibrator have modes, associated to resonance frequencies and their harmonics. This translates in the observations in a "comb of lines", characteristic of molecular lines (fig. 2.13).

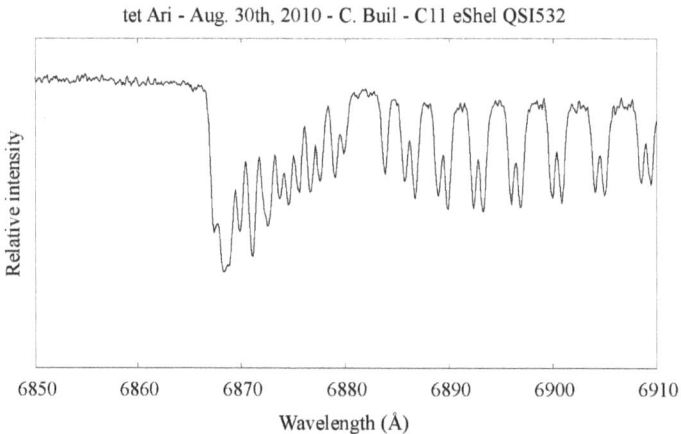

tet Ari - Aug. 30th, 2010 - C. Buil - C11 eShel QSI532

FIG. 2.13 – Spectrum of molecular lines.

Chapter 3

What Light Tells us about Stars

After reviewing the nature of light and the different methods to make it, I wish now to "connect" this information to astronomical objects. Keeping an eye on the orders of magnitudes.

I pay homage to the marvelous "gift" we receive from the sky. We saw that many phenomena can make or slightly modify light. Once this light is made and shot into space, it can travel for *millions of years* without the smallest change. Figure this amazing fact: millions of years – and the light reaches our telescopes intact. Any micro-phenomenon would suffice to modify just a tiny bit the light, and we would not be able to do anything with our observations.

Another marvelous element essential to do astronomical spectroscopy is that the physics in the stars is the same as we know on Earth. Of course, the environments are different (temperature, pressure, chemical composition, etc.). We can't really check exactly...but the nature of the spectra observed from stars matches perfectly with what we observe on Earth. God would be very teasing to make such a perfect illusion!

If these two conditions (same physics throughout the Universe, and unaltered light) were not met, this book would not have a subject, spectroscopic instruments would not give any useful information, and we would not understand anything about the Universe surrounding us. I like this humility imposed by Astrophysics : there are things that go far beyond us.

Since most of the astronomical objects shining light toward us are stars, it is useful to start with a – very brief – summary of how a star works.

3.1   The Light of an Ordinary Star

A star is a giant ball of gas, so compressed by its own gravity that nuclear fusion reactions ignite in its center. These reactions start when the core temperature reaches about 10 million degrees, and they mostly transform the hydrogen in the core into helium. The star finds itself in hydrostatic equilibrium, with the energy released by the nuclear fusion balancing the compression from gravity.

FIG. 3.1 – 2D spectrum of the Sun.

The energy produced leaves the star as an intense light. Depending on the star mass, the surface temperature of the star will be roughly between 3,000 K and 50,000 K. It is the light from the stellar surface that is sent into space and reaches us.

Astronomers defined spectral types to classify the stars, indicated by letters: O, B, A, F, G, K, M. Actually, the spectral type corresponds to the surface temperature of the star, with the O type stars being the hottest, and M the coolest.

Thus, the first thing we see in a stellar spectrum is the black body profile corresponding to the surface temperature of the star.

Before leaving the star, it takes a very long time for this light to cross the inner layers of the star, and in the end, it is filtered by the chemical elements in the surface layers. The main elements are hydrogen and helium, but we also find many other atomic species. By filtering the light, these elements create many absorption lines in the spectrum.

Therefore, a stellar spectrum is very different from the spectrum of any artificial light source on Earth: it is the combination of a Blackbody profile and many absorption lines.

For example, the light from the Sun produces the spectrum presented in figure 3.1 (in the visible domain).

Recall that a Blackbody profile has its maximum intensity in the visible range for temperatures between 3,800 K and 7,600 K. This temperature range covers many existing stars, thus we can find many different spectra. For example, the graphic (fig. 3.2 shows three Planck's profiles for 3,000 K, 6,000 K and 30,000 K for the frequency range of interest (between 400 nm and 700 nm). A hot star has a globally decreasing profile from blue to red, while a cool star has a globally increasing profile.

Temperature is probably the first measurement you will have the chance to make in spectroscopy: by putting a Planck profile on top of your spectrum and playing with the temperature parameter to make them match as close as possible you can define a star's temperature with great precision.

Black body profiles

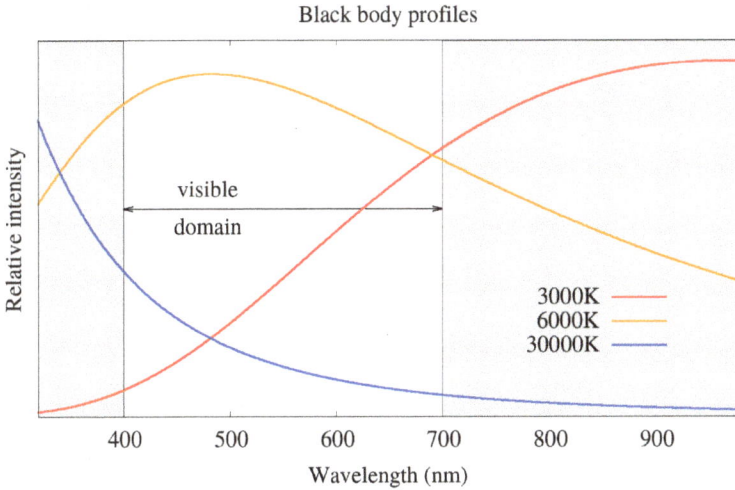

FIG. 3.2 – Planck's profile for different temperatures. Note : the vertical scale of the curves has been adjusted for the sake of readability.

Let me digress for a moment. It is a continuous amazement for me that to the naked eye, all stars look almost the same – they are just shiny dots. Some say we can see some colors (I confirm!), but only if we look very carefully we can distinguish different shades. As soon as we do spectroscopy, the differences can be easily seen. There is no way of missing it. It is not about searching for fine details. The profile of a hot star looks *nothing* like the profile of a cool star. With a little bit of practice, you will be able to easily discern the profile of a B star from the profile of a K star.

This is even easier since depending on the stellar type – that is, the stellar surface temperature – the atoms are excited differently. The result is easily seen: a hot star has a very smooth profile, very close to the Planck profile (it is easy to measure its temperature), with only a few lines (essentially from hydrogen). Figure 3.3 shows the spectrum of the star *I Leo*, type A2V, as an example.

In contrast, a cool star has a profile full of lines, to the point that it is hard to see the Planck profile. Figure 3.4 shows the spectrum of the star *HR5086*, type K5V.

3.2   Each Star has its Own Spectrum

I have briefly described the spectrum of an "ordinary" star, i.e. a star on the main sequence of the Hertzsprung-Russell diagram (HR diagram) – where stars spend most of their life. This diagram is useful to describe the evolution of a star throughout its life.

I Leo - Apr. 18th, 2015 - F. Cochard - Alpy 600 C8 Atik 314L+

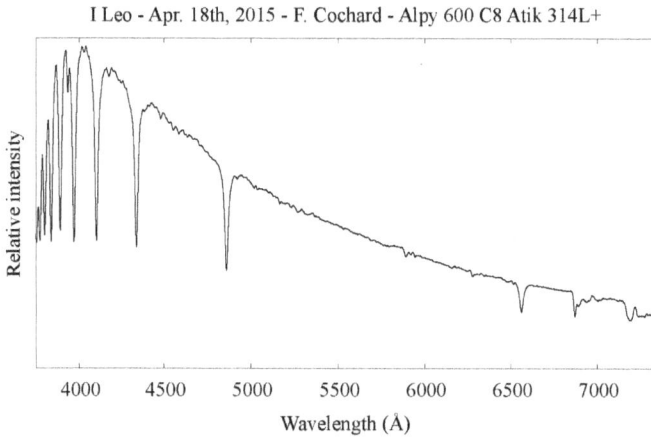

FIG. 3.3 – Spectral profile of a hot star.

HR5086 - Apr. 17th, 2015 - F. Cochard - Alpy 600 C8 Atik 314L+

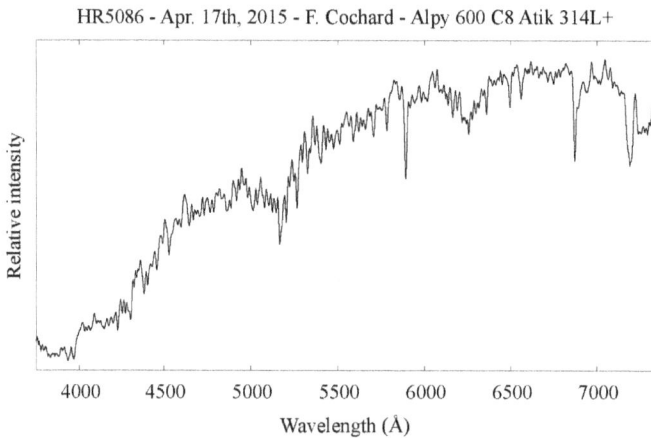

FIG. 3.4 – Spectral profile of a cool star.

Reality is more complicated – there aren't two stars in the sky with the same spectrum – and many parameters (other than the temperature) determine the immense variety of stellar spectra:

– the stellar mass;

– the environment in which each object is born varies, therefore the chemical elements in the stellar envelopes are specific to each star;

– the advancement on the HR diagram also leads to very important variations in the spectra.

Be star HD26398 - 2013-12-15T20:13:55 - F. Cochard - Alpy 600 C8 Atik 314L+

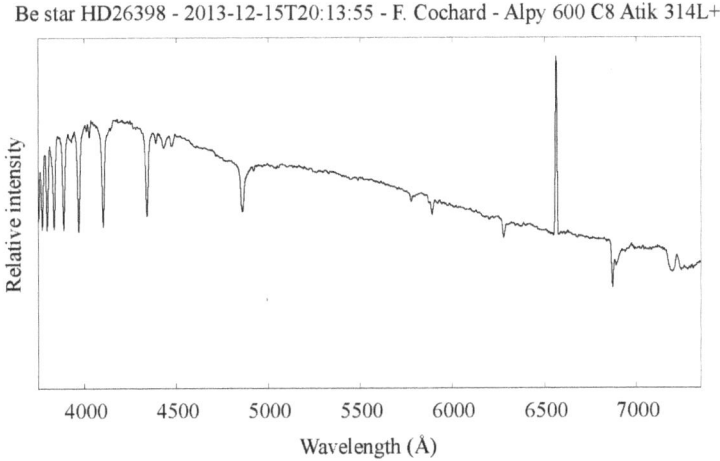

FIG. 3.5 – Spectral profile of a Be star

Moreover, many different types of stars show emission lines: Be stars (fig. 3.5), Wolf-Rayet (WR) (fig. 3.6), planetary nebulae (fig. 3.7), novae (fig. 3.8), etc.

3.3 Observing at Different Resolutions

It is useful to stop and think about what we can effectively observe in these stellar spectra. Until now, I have essentially talked about spectra covering the entire visible range. But depending on the spectroscope we use, we can observe the spectrum for the whole visible range, or we can focus on a small portion of it. We can use an analogy with photography: in front of the same scene, we can choose objectives with different focal length to see a wider view or more narrow details of this scene.

We need to talk about the instrument resolution, that is the smallest detail discernible in a spectrum obtained with it. The higher the resolution of the spectroscope, the larger the linear dispersion of the produced spectrum. Obviously, there is a compromise between the resolution and the spectral domain observed, simply because of the size of the detector. The more we "zoom in" on the spectrum, the smaller the spectral range that is covered.

Here are some orders of magnitude for instruments commercially available today. We talk about low resolution for spectra covering the whole visible range (possibly extended to the near infrared). Therefore, the extent of the covered domain is of order 350 nm (between 400 nm and 750 nm), and we access details of order 1 nm (this is the case of the Alpy 600, for example). With high resolution, we can cover a domain of order 10 nm, with details of

WR154 - 2012-06-28T23:21:23 - C. Buil - C11 LISA

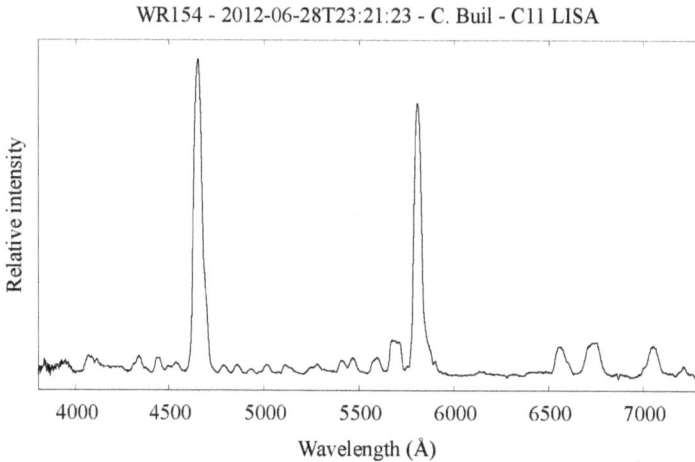

FIG. 3.6 – Spectral profile of a WR star

Planetary Nebula M76 - 2011-09-27T23:14:35 - C. Buil - C11 LISA

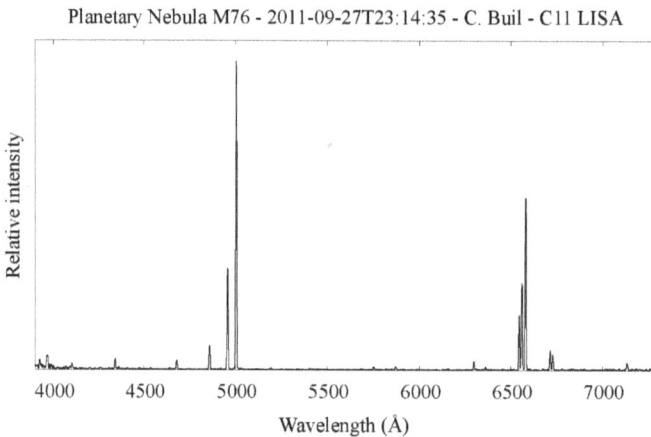

FIG. 3.7 – Spectral profile of a planetary nebula

the order of 0.03 nm (as for the Lhires III, for example). Thus, there is a ratio of 1 to 30 between this two kind of instruments, and you will need a series of thirty high-resolution spectra taken one after the other to cover the same domain as in the low-resolution case.

Let's consider the Be star *β Lyr* as an example (it is a star close to Vega, in the Lyra constellation, commonly called *Sheliak*). At low resolution (figure 3.9, obtained with an Alpy 600 – $R = 600$), we see the whole visible spectrum. We clearly see the $H\alpha$ emission line (6563 Angstrom) in the red portion of the spectrum, but we cannot distinguish any further details in this line.

Nova Sgr2014 - 2014-03-02T05:28:09 - C. Buil - C11 Alpy 600

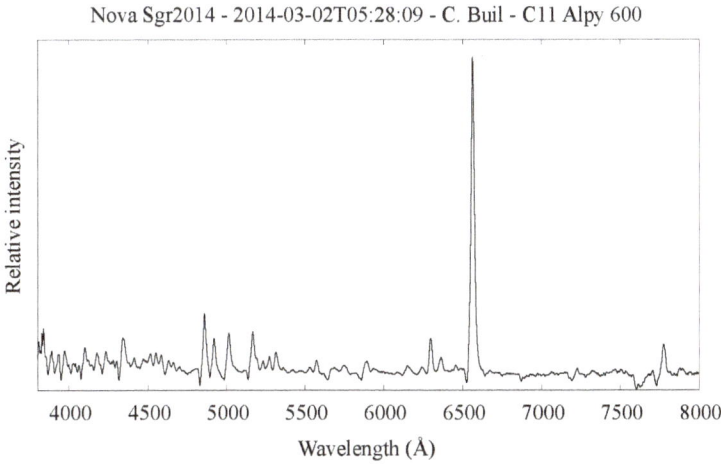

FIG. 3.8 – Spectral profile of a nova

bet Lyr - 2015-02-28T06:02:11 - C. Buil - T200 Alpy 600

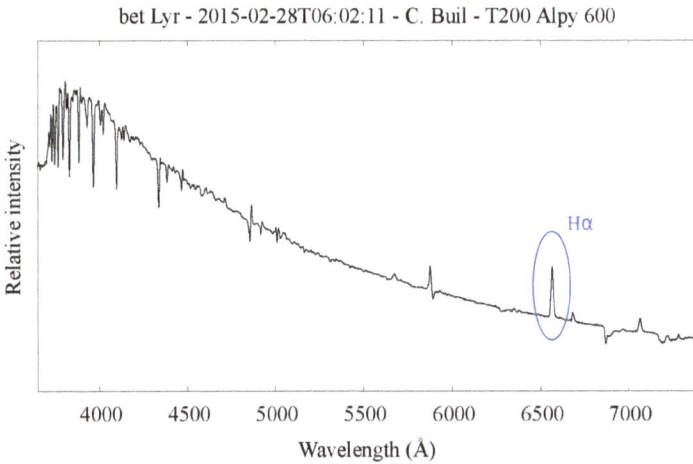

FIG. 3.9 – Low resolution ($R = 600$) spectrum of β Lyr.

If we now observe the same star at high resolution (figure 3.10, obtained with a Lhires III – $R = 18,000$) around $H\alpha$, we can see a plethora of details.

When we work at high resolution, we have the problem of choosing the spectral domain to observe. If we have a spectral range of 10 nm, we can choose to put it in the blue, green of red range.

Generally, if we observe at high resolution, it is to focus on a very specific line (either absorption or emission). We look for the detailed profile of the line, which can itself contain a great richness of information.

Since hydrogen is the main constituent of the universe, it is also the main element in stars. Generally, it is the element with the strongest lines. In the

bet Lyr - 2014-06-26T00:56:12 - A. Favaro - C8 Lhires III 2400

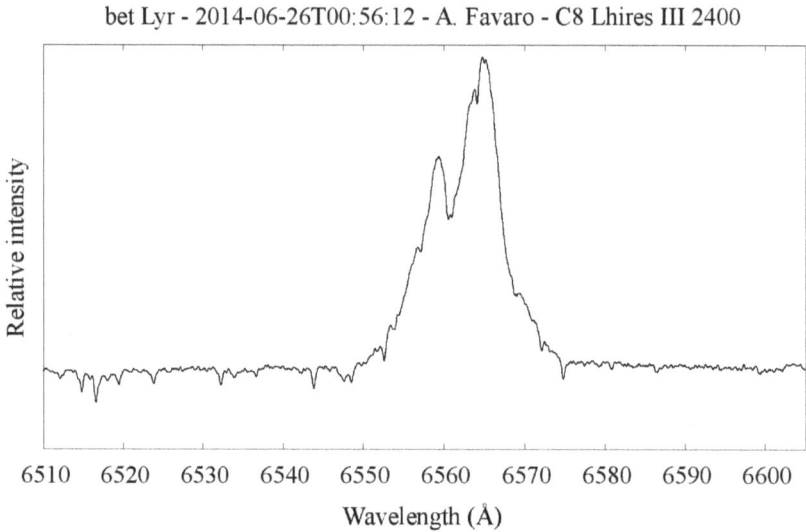

FIG. 3.10 – High resolution ($R = 18,000$) spectrum of β Lyr, centered on the line $H\alpha$.

visible domain, we see a series of hydrogen lines called Balmer lines; the most important of these lines is $H\alpha$ (pronounced H-alpha) at 656.3 nm. In practice, if you don't know where to start from, look at $H\alpha$: it is *in some sense* the common denominator of all stars. If you have the occasion to visually look in a spectroscope (whatever its resolution), this line is at the edge of the visible spectrum, in the deep red.

We saw that the spectrum is modified by the Doppler effect, and that the radial motion of the observed star can shift the spectrum by about 0.1 nm per 50 km s^{-1}.

If a spectroscope is capable to distinguish details of 1 nm in a spectrum, we can consider it is capable of measuring the position of a line to a tenth of this value – that is 0.1 nm.

Let's apply these values to the resolutions below:

– at low resolution, we can see details of order 1 nm, therefore we can measure velocities of order 50 km s^{-1};

– at high resolution, details of order 0.03 nm allow the detection of velocities of order 1.5 km s^{-1}.

In any case, these are very high velocities compared to our everyday life. But in space, these are very common.

Let me mention here a very important element about the physical measurement of radial velocities. The numbers above are for a specific single line. But if we have a spectrum with more lines – as it is often true when we cover a large spectral range – we can use a very powerful numerical method called Cross Correlation Function, or CCF. CCF strongly enhances the sensitivity in radial velocity by at least a factor of 10.

3.4 Limiting Magnitude

As soon as we talk about resolution, we need to discuss the limiting magnitude that we can observe. The magnitude measures the brightness of a star. It is higher the fainter the star is, and follows a logarithmic scale. The magnitude decreases by one unit when the brightness is divided by roughly 2.5 times. A difference of 5 magnitudes between two stars corresponds to a change in brightness of 100. The brightest stars in the sky (except the Sun) have a magnitude of about 0, and the faintest stars observable with the naked eye with a very clear and dark sky have a magnitude of about 6. The faintest stars ever observed with the largest professional telescopes have a magnitude lower than 30.

In contrast with imaging, when all the light from a star is focused on a few pixels of the CCD detector, spectroscopy largely spreads this light – and the higher the resolution, the less light each pixel receives. In the case of high resolution, we only take a small portion of the energy in the spectrum. Up to a certain point, increasing the exposure time can compensate for this.

With an amateur instrument, for example a 200 mm ($\sim$ 8 inch) diameter telescope, we can easily access magnitude 10 to 12 with a low resolution spectroscope and a few minutes of exposure. At high resolution, instead, we need something like an hour of exposition to reach magnitude 7 to 8. These numbers are indicative and generally achievable in practice: I know amateurs that have gone far beyond, having taken particular care in their observations.

Through these few numbers, you see that we need to find a compromise between resolution, magnitude and exposure time. Thus, there is *no ideal spectroscope* : the best choice depends on the given observations, and on the scientific goal you are trying to achieve.

Another way of saying this is that high and low resolution are very complimentary: low resolution allows one to observe faint objects, or more ordinary objects with shorter exposure times (therefore allowing for many observations in one night). High resolution allows one to see the fine details in a spectrum.

There are certain cases in which we want a large spectral range, high resolution, and short exposure times all at the same time... up to a certain point, the two first points can be achieved (with Echelle spectroscopes – see section 5.10), but the only real solution to observe faint magnitudes, without increasing crazily the exposure time, is to increase the size of the telescope (together with the size of the spectroscope). This is why professional astronomers are running for gigantism. Nevertheless, I can assure you that the number of stars in the sky is such that, even with a modest instrument, you can have fun for many years.

I have to underline here a few particular cases. The numbers I gave here are for "normal" stars, with spectra following roughly the Planck profile. But there are objects for which this is not true. For example a nova is an exploding dying star and it shows a sudden brightness increase. It turns out that the increase is generally centered around the $H\alpha$ line – at least in certain phases of the explosion. Thus, the visual magnitude of a star can be misleading, since it measures the energy over the entire spectrum. If you only observe this region of the spectrum (around $H\alpha$), at high resolution, the brightness will be much higher than what the magnitude of the star suggests.

3.5   A Moving Sky

When we observe the sky, with the naked eye, or with a small instrument, we quickly convince ourselves that nothing moves, except the rotation of the sky, the motion of the Sun, and the motion of planets. If we use a slightly bigger instrument, with a CCD camera, we can see the motion of asteroids. All these phenomena are restritcted to the solar system, which is our closest neighborhood (on the Universe scale). Only a few close stars show apparent motion in the sky visible on a human scale, for example because of parallax while the Earth moves along its orbit around the Sun (for example Barnard's star). Other than this, everything is perfectly still and motionless... Or is it?

Of course not: The universe is in perpetual motion. If we can't detect it with our senses, it is just because the speeds involved are much too small compared to the distances to the stars.

Amateur spectroscopy allows for the measurement of intrinsic displacement velocities, no matter what the distance to the source is. It does not allow for the measurement of the distance (directly at least), but it gives access to velocities of a few ten km s^{-1}, or even less.

Let's have a look at a few example of motion we can observe with this range of velocities.

It seems appropriate to start from the velocity of Earth on its orbit around the Sun. Earth is about 150 million kilometers from the Sun, and goes around it in 365 days. A quick calculation yields an orbital velocity of Earth on its orbit of about 30 km s^{-1}. This means that if we observe the same star with a six months interval, once Earth is approaching the star at 30 km s^{-1} (if the target star is on the ecliptic, otherwise projection effects change this value), and once it is moving away at the same speed. Therefore, there is a difference of 60 km s^{-1} between the two measurements: this is easy to detect with high resolution amateur instruments (fig. 3.11).

Let me emphasize this first example of radial velocity: with a very modest instrument, we can measure the displacement velocity of Earth on its orbit around the Sun. Starting from this velocity, and the observation of the annual period of the sky movement, we can work out the Earth-Sun distance, and

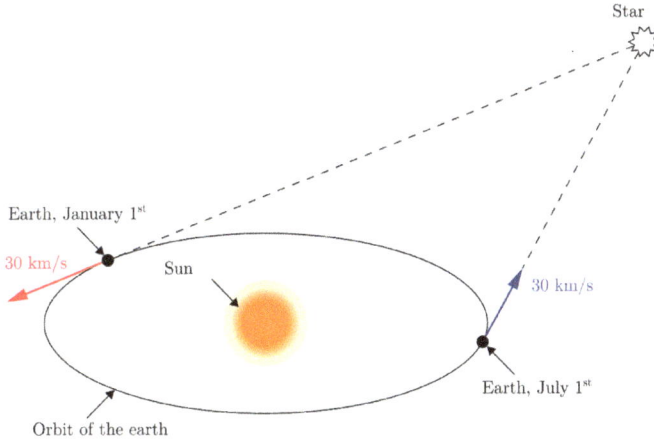

FIG. 3.11 – Motion of the Earth around the Sun

from this, using Kepler's third law, we can measure the mass of the Sun. Measuring the mass of the Sun from your backyard...pretty cool, isn't it?

This has also a surprising consequence: as soon as you observe a star of which you want to know the intrinsic velocity, you have to correct your measurements for Earth velocity with respect to the star. This velocity is called *heliocentric velocity*. It depends on the exact date, terrestrial coordinates of the observing site, and celestial coordinates of the observed object. Therefore, it is absolutely crucial to note this information at the time of observing (it is almost automatic with most software, I'll come back to this).

Then, there is the proper radial motion of stars in the Galaxy. These are generally quite small, but most are detectable: they span the range between a few ten km s^{-1} to roughly 100 km s^{-1} (for the fastest stars) relative to the Sun. It is an easy measurement to do, especially on the fastest, and very instructive.

Other than the proper radial velocity of stars in the galaxy, there is another phenomenon that is particularly impressive. A large fraction of stars (roughly half) are binaries, that is, they are really two stars orbiting each other – or, to be more precise, orbiting a common center of mass. During this revolution, each star shows a periodic oscillation seen from Earth. It approaches and then moves away, in phase opposition with the companion.

Most of binary stars are not visually separated – we see one dot, even with the largest telescope in the world – but their spectrum is the superposition of the spectrum of each star (and these can obviously be of different spectral types). And since each star in its orbit shows a relative motion, we see the splitting of the absorption lines in the combined profile. It is this splitting that shows a periodic variation.

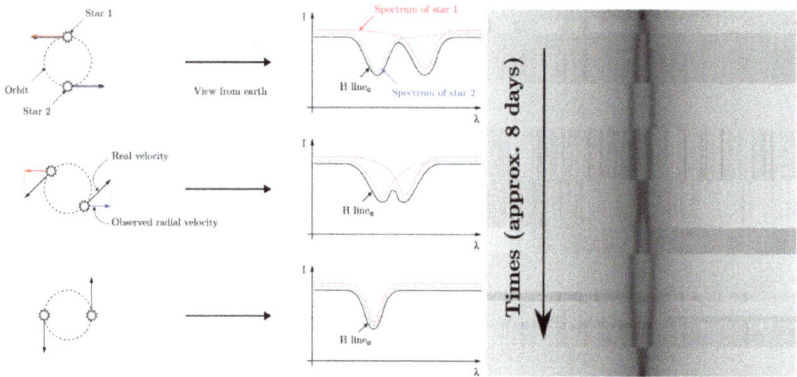

FIG. 3.12 – Dance of the spectra of β Auriga (the right panel shows a real sequence of observations spanning roughly 8 days)

The time scale for this motion can be extremely variable: from a few days to decades. The latter requires a lot of patience, but the former are really spectacular to observe. The amplitude of the velocities spans a wide range (for a given stellar mass, the orbital velocity and the period are linked by the Kepler's third law). Velocities of a few km s^{-1} are common for binaries with short periods (less than a few weeks): therefore, it is a phenomenon easily accessible to (high resolution) amateur instruments. For example, the star β *Auriga* has a magnitude of about 2, a maximum radial velocity between the two stars of about 200 km s^{-1} and a period of 4 days: a few nights are enough to see the dance of their spectra (fig. 3.12).

Since the two stars orbit around their common center of mass, they don't necessarily have the same velocities: the less massive will have a larger orbit, and since the period is the same, also its velocity will be greater. This means that if we measure a difference in the velocities of the two stars, we have a direct measurement of their mass ratio. And, like for the Sun, period and velocity give us the distance between the two stars, and therefore the two masses of the stars. There is more than just the mass of the Sun that we can measure from our backyards!

This is exactly the same technique used for the detection of exoplanets – but there is another order of magnitude here, since the largest velocities are about 200 m s^{-1}, and Earth-like planets induce on their star an oscillation of a few cms. The fastest exoplanets (including *51 Peg*, the first discovered back in 1995 at the Observatoire de Haute-Provence) are now beginning to be accessible to amateur astronomers[18]...I bet it's just the beginning.

If a star *really* were an immobile point, most absorption or emission lines would be quite narrow. This is not the case, and in practice, with high

---

[18] See the measurements from Christian Buil : http://www.spectro-aras.com/forum/viewtopic.php?p=4104

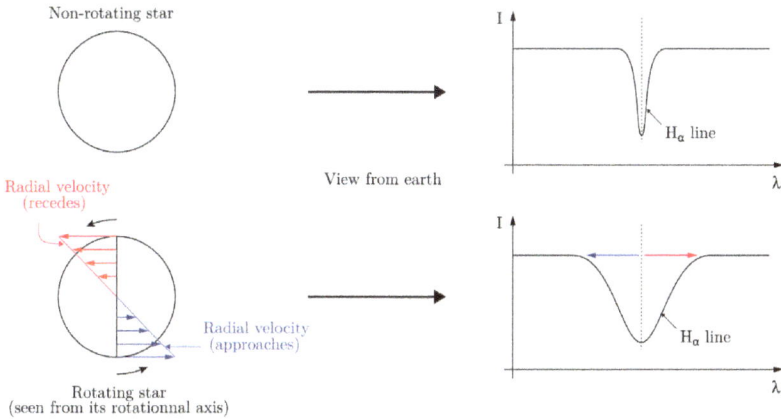

FIG. 3.13 – Stellar rotation and line broadening.

resolution we observe that lines have a non-negligible width. The broadening of lines is the result of the thermal agitation within the star (linked to the temperature, and intrinsic to the source) and of the rotation of the star on its own axis (which depends on the observers frame, fig. 3.13). These phenomena create a relative velocity of the source, which translates to a Doppler effect for the observer, and thus a shift in the spectrum. The result of the distribution of velocities within the star is line broadening.

Sometimes, it is even possible to distinguish different profiles within a single line, which reveal the multiple physical phenomena entering into the line formation process. For example, if you observe *Vega* – one of the brightest stars in the summer sky – at high resolution (around the H_α line), you can clearly see a break in the lines profiles (see fig. 3.14, top panel): it is intuitive to think that two different processes determine these profiles (fig. 3.14, bottom panel). These profiles are named *Voigt Profiles*.

I refer you to the literature to further explore what other information spectral line profiles contain.

Let me digress briefly on another phenomenon which is not yet accessible to our instruments, but by so little that I would not be surprised if it will be in a few years. A star is a mass of gas, and as any other physical object, it has resonant modes, therefore there will be pulsations at its surface. These pulsations translate into local additional components of the velocity, which can be detected with very precise instruments. The first measurements of this kind have been done on the Sun (so-called helioseismology), but we have now the capability of similar measurements on other stars – this is called asteroseismology.

Let's get out of the galaxy now.

When we deal with observations of galaxies, one question often comes up: can we measure the cosmological *redshift* and recover the Hubble law (which describes the expansion of the universe)? The answer is *yes*, up to a certain

Vega - 15-03-2014 - C. Buil - C11 Vhires (R=48000) - Castanet

Vega - 15-03-2014 - C. Buil - C11 Vhires (R=48000) - Castanet

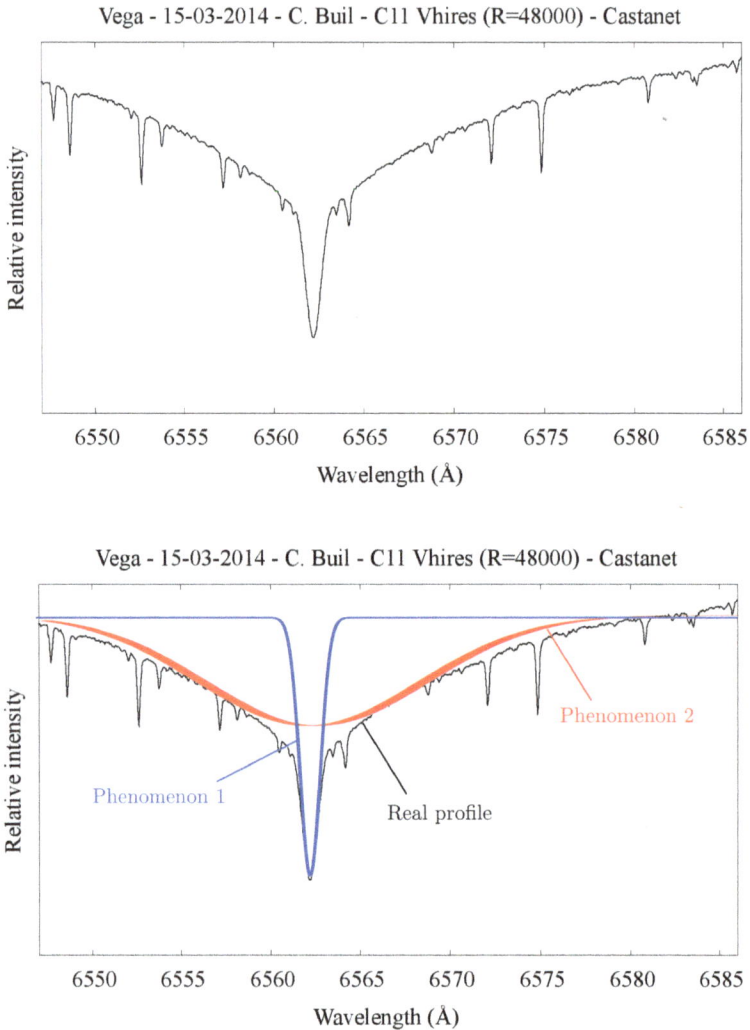

FIG. 3.14 – High resolution spectrum of *Vega* centered around H_α (Voigt Profile).

point. The farther the galaxies are, the faster they move, the larger (and thus easier to detect) the Doppler shift is, the more complicated the observation is because of the limiting magnitude. The recession velocity of galaxies ranges from a few hundreds to several thousands of km.s^{-1}.

Let me emphasize that galaxies are groups of a very large number of stars. What we observe is the "averaged" spectrum of all these stars. Moreover, galaxies are diffuse objects, with a surface visible in the telescope.

But there are even more exotic objects. For example, quasars, which deserve a special comment: they are extremely bright objects in very far galaxies (the galaxies themselves are not visible). They have a large recession velocity and their spectra are vastly shifted. The most famous case is the quasar $3C273$: its velocity is of order 45,000 km.s^{-1}. It is an almost mythical object.

3.6   Time Evolution

When I previously described the spectrum of an ordinary star, I implicitly assumed that the star is static, and immobile. Indeed, it takes several million years for an ordinary star (for example of one Solar mass) to leave the main sequence: the intrinsic evolution of its spectrum is totally imperceptible on a human timescale...and even on a timescale of several generations.

And yet, not all stars are "well behaved", and many celestial phenomena evolve much faster; once again the range of timescales is very wide. We can classify these phenomena of variability into two broad categories: periodic and random phenomena. Periodic phenomena – for example binary stars, mentioned in this chapter – are predictable, at least to the first order. This vastly simplifies the observations.

Another example is pulsating stars: these unstable stars have intense and fast movements of their envelopes. For example, the star *BW Vul* shows pulsations with velocities of order 200 km s^{-1} with a period of few hours – the energy involved is absolutely huge.

The random phenomena, or *cataclysmic* (this name comes from the incredible energies involved), are extremely important in Astrophysics: there is more physics to unveil in a star changing abruptly, than in a star in a very stable equilibrium. In reality, except for "catalogue surveys" (which consist of an inventory of all objects of a certain kind, or in a certain region of the sky), most of the observational attention is dedicated to the phenomena of variability, on short, or even random, timescales.

The random phenomena are very numerous. For example, Supernovae (i.e. the explosion of a massive star ending its life, visible up to very distant galaxies), can happen at any time, and they last from a few days to a few months. Generally, these phenomena are discovered in photometric observations – I've heard that you need to observe something like 1,000 galaxies to find a supernova. Some automatic surveys scan many hundreds of galaxies each night...there is never a lack of supernovae to observe.

Novae are also part of this category of cataclysmic stars. They are also stars that appear suddenly, but this time, they are stars within our galaxy. The explosion that triggers the increase in brightness is tremendous, but nevertheless insufficient to disrupt the star. After this violent outburst phase, the star goes back to its quiescent stage for years (sometimes centuries) and can then outburst again.

3.7   Not only Stars

Until now, I have only dealt with observations of stars, since they are the origin of most of the light we can detect from the sky. However, there are also other objects well worth some of your observation time.

Among the easiest objects to observe are bodies in the Solar system: the Moon, planets, asteroids, and comets – not to mention the Sun itself.

Except the Sun, these objects do not emit light on their own, but just reflect sunlight. First of all, we observe the solar spectrum in these objects, but modified by the nature of their composition. Thus, we typically do not look for the spectrum itself, but for the differences with the initial solar spectrum.

Comets are very cold objects, which approach temporarily the Sun. The Sun heats their surface layers, which evaporate and form the comet tail. Because of the Sun, the elements in the tail can be excited, and then produce their emission lines. The spectroscopic observation of a comet's tail allows one to access the chemical composition of its external layers.

Farther from us, in the heart of the galalxy, we find many nebulae, generally presented as *stellar nurseries*. These are giant clouds of gas and/or dust, which can produce emission lines – like the comets – if there are stars to excite this gas.

If the cloud is not excited by nearby stars, it remains inaccessible to our instruments – we then talk about interstellar medium. On the other hand, the light from stars in the background has to cross it to reach us. This translates into a modification of the spectrum, typically a *reddening*: the blue portion of the spectrum is absorbed more than the red one, and the global profile of the star departs from a Planck profile. Moreover, we see extra absorption lines in the stellar spectra, caused by these clouds. These are typically very thin lines, since the thermal motion in the insterstellar medium is very weak.

3.8   Basic Chemistry

There is a major difference between the stars and non-stellar objects: stars are so hot that almost no molecule can exist in it. The atoms are free and stellar spectra are atomic spectra. It is the degree zero of chemistry!

On the other hand, all the other objects I mentioned (planets, asteroids, comets, interstellar medium) are extremely cold, and in these environments atoms can group into molecules to form larger and larger dust grains.

As always in astronomy, there are notable exceptions. Nebulae, for instance, are gas clouds that can be heated up by nearby stars: their temperature can be higher than the temperature of some stellar atmospheres!

We have seen in section 2.6 that the spectra of molecules are richer in the infrared than in the visible. This is why the observation of comets or interstellar dust clouds is preferentially in the infrared.

Astrochemistry is still very rudimentary compared to terrestrial chemistry. The molecules we can find in astrophysics are very simple: in a certain way, the chemistry of astronomers stops where the one of chemists starts.

Chapter 4

What can I Observe
with my Instrument?

In the previous chapter, I have tried to show every information light can give us about celestial objects. We have seen briefly the diversity of physical phenomena we can observe, and the nature of objects in the Universe. We are now going to deal with how to transform all this potential into very real observations. What is accessible to my instrument and my observational conditions? What are the difficulties I should expect?

The aim here is to define which type of instrument to choose for a given observation, or to adapt your observations to your instrument, if you already have one.

4.1   The Basic Questions

Before any observation, it is wise to ask yourself some questions:

- What is the targeted object magnitude?

- What is the resolution I need?

- What spectral domain do I need to cover?

- What is the required exposure time?

- What are the intrinsic constraints of this observation?

Magnitude

The brighter an object (the lower its magnitude), the easier it is to observe. The low magnitude makes it easy to find and follow the object, and the exposure times are shorter. Always chose bright objects to start, then, when you master your topic, move on to explore fainter magnitudes.

For an amateur astronomer with a small instrument (for example a 200 mm-diameter telescope), an object can be considered bright (thus easy) down to magnitude 4-5. Beyond this, it depends on the resolution (see below). To first order, we can go down to magnitude 7 at high resolution, and to 12-13 at low resolution, with moderate effort. It is possible to go even beyond, but only with particular conditions (clear sky, perfect mastering of the instrumentation, etc.)

Keep in mind that in increasing one magnitude, the luminosity decreases by a factor of 2.5...it goes very fast.

Resolution

The resolution determines the capability of your instrument to see details in a spectrum. We often use the notion of *resolving power* to characterize an instrument:

$$R = \frac{\lambda}{\Delta\lambda}$$

where λ is the wavelength considered, and $\Delta\lambda$ is the smallest detail visible in the spectrum. I will come back to this when dealing with the instrument.

Depending on the purpose of the observations, you can look for a *wide view* (especially at low resolution), to cover the entire visible spectrum. This is the case, for example, if you explore the different spectral types, or to determine the effective temperature of a star. On the other hand, if you want to work on radial velocities, or if you want to distinguish two lines very close to each other, you need a high resolution. But keep in mind that the higher the resolution, the more dispersed the light, the longer the exposure time; and on top of this, the spectral range decreases. In the amateur community, we commonly work with resolving power in between $R = 100$ (low resolution) and $R = 20,000$ (high resolution). Professional instruments often fall within the same range, but in some special cases they can reach very high resolution – $R = 100,000$ or more.

Spectral Domain

The question of the spectral domain concerns mostly high-resolution observations - since the spectrum is more dispersed, and CCD detectors have limited range. In these cases, you need to define which portion of the spectrum you want to observe. If you need to observe a large region which does not fit in your detector, you can take a series of spectra with partial overlap, to later make a longer spectrum – the same way you can take a panorama with a camera. It is possible, but rather annoying. It can also happen that an observation requires two disconnected domains – for example the lines H_α (at 656 nm) and H_β (at 486 nm). In these cases, instead of covering the whole range with a single instrument, you can make two consecutive, and independent, observations.

In the end, you always have to ask yourself: "what is the spectral domain I need to observe, and is my instrument capable of doing it?".

Exposure Time

Of course, the fainter the object you target, the longer the exposure time. This is also true for the resolution: the higher the resolution, the more dispersed is the light, and the less light each pixel receives. In our context, the typical exposure time ranges from a few seconds for bright objects, to an hour for a high-resolution observation of a faint object (magnitude 7 for example). Since astrophysical objects are generally very stable, you can expose for hours in a row, to compensate for the small amount of light available. Of course, there are limits to this:

- long observations are annoying. When you start a two hour observation, times slows down...especially at two in the morning!

- a corollary of long exposure time is that the number of observations you can realize in one night is small. It takes a lot of patience to observe for a whole night and only get a few stellar spectra;

- Some phenomena evolve very fast. Pulsating stars such as *RR Lyrae* change in a few minutes during the critical moments. A long exposure time is no use in this situation because you would miss the interesting phenomenon.

Intrinsic Constraints

Many observations can be done at any time: on the timescale of our entire life, the phenomenon does not change a bit. But in amateur spectroscopy, we like to focus on rapidly changing things – precisely for this show. Typically, there are constraints on the best observational window for rapidly varying objects. Let's cite again the example of *RR Lyrae* stars: these are pulsating stars (period of a few hours to a few days), but the best moment to observe them is at their maximum. For a good observation, we need the maximum to be during the night... which means not all the maxima are observable. And stars seldom care about our everyday life: job, week-end, meteorologic conditions... this quickly gets in the way of observations.

Even without going for such complicated observations, keep in mind that only circumpolar stars are observable all year long from the northern hemisphere. Many stars are visible only in a fraction of the year, and a large portion is observable only from the southern hemisphere.

4.2   Many Types of Observations

Your first step will probably be the observation of a few bright and isolated stars. But while you get more practice, you will soon be more ambitious in your observational program. Everybody has their own character and constraints, but it seems useful to give you a few tips on the observations regularly done in the amateur community. This can allow one to determine whether this or that program is at your level, or suits your interests. We can define several observational programs:

- isolated observations;

- long term observations;

- observational campaigns;

- follow-up observations;

- events.

This classification is suitable for leisure or educational observations (understanding the sky, improving techniques), and for observations with a strong scientific ambition (for example in Pro-Am collaborations).

Isolated Observations

This is probably where you want to start from: observe simple objects with no particular time constraints. If you fail the first night, try again the next one: the star is still there. This is also the reason why they become boring quite soon: they are desperately predictable, without surprises.

Since they are very stable, they are ideal to compare your results with colleagues; very useful to monitor progress.

Long Term Observations

As soon as we have sufficient skills, it is helpful and fun to add a temporal component to your observations. We saw that the sky offers a large variety of time scales. The observation of a binary star revolution for instance, or the Earth around the Sun, require frequent observations, say once a month. This requires regular observations, bringing out all the instrumentation...but it is very gratifying to say "I saw the sky moving"!

Observational Campaigns

Regularly, groups of observers organize observations in the context of larger campaigns (Pro-Am). This can happen for particular events. For example, in 2011, the star δSco was in a very important phase: it is a recently discovered binary Be star (with period of 10 years), and the two components were at their closest (periastron) then. Researchers were expecting a great increase in activity, easily observable with amateur instruments. The observations have been numerous, and the result largely satisfies the expectations: we have seen some real fireworks. In the end, it was established that the star has probably a third companion, which explains the high instability in the system.

This can also happen to focus the observations in a given time frame (to have a good time cadence in a finite time period), or to help observations with large professional instruments (amateurs can provide intermediate observations), or even to provide triggers (in the δSco case, observers were in charge of detecting the beginning of the activity in 2011).

Follow-up Observations

Another kind of observational program consists in regularly observing a certain group of stars, either to detect random phenomena, or to characterize a stellar population. This is the essence of the BeSS program, a very well structured Pro-Am collaboration lasting for more than a decade now. Its aim is to regularly follow a few hundred Be stars to detect the largest number of transitory phenomena (so-called *outbursts*), and to possibly find some periodicity in these stars. These are "long term programs", not always spectacular on the short term, but extremely useful to research on a longer timescale.

The Events

Frequently, stars explode as novae or supernovae, and amateurs' observations are extremely useful in these cases. They are also very rewarding: almost everyday there are scoops! Most of the time, the discoveries are made photometrically (by amateurs). At this moment, it's urgent to observe in spectroscopy.

The Nova that appeared in the Dolphin constellation in 2013 (nova Del 2013) will be remembered as the turning point in Pro-Am collaborations: no nova had been observed with such high temporal cadence before. No professional observatory could have possibly done such a campaign – the first spectrum was taken less than two hours after the announcement of the discovery by Olivier Garde. The data[19] from this campaign will certainly keep the researchers busy for a few years.

These event observations are spectacular (the spectra change a lot within an hour), and they are a lot of fun: they are clearly an activity on the rise.

Fishing for Anomalies

Let me mention another kind of program, seldom exploited but which would also deserve developments: we can look at a given sample of stars and

---

[19] See the Aras website http://www.astrosurf.com/aras/Aras_DataBase/Novae/Nova-Del-2013_2.htm

look for any kind of anomaly. By definition, you never know what you could find – as a fisherman throwing in his line. You need to be a gambler, because it is probable that you will never find anything. But by proceeding like this, you also risk making a discovery nobody has ever predicted. It is a bit like the lottery: few chances of winning, but if you do, your life is changed.

Among these potential programs, some are less random than others. I give you an example I deeply care about: the discovery of new Be stars. A Be star is a type B star (hot, and active) for which a Balmer line (typically H_α) has been seen in emission at least once. More than 2,000 Be stars are presently known, and these are probably all the Be stars down to magnitude 8 or 9 – roughly 20% of B stars. But beyond this, the ratio of Be to classic B stars drops – and there is no apparent reason for that, since the magnitude depends only on the distance to Earth. It is most likely because a B star of magnitude fainter than 9 have been seldom observed by humanity... what a beautiful playground for amateur observers today! It is probable that if you observe B stars with magnitude fainter than 9, you will find new Be stars.

4.3   What Physical Phenomena to Observe?

Another way of looking at things is to ask which physical phenomenon do we want to observe. For example, if you want to measure a radial velocity, most likely you will need high resolution. Recall that a velocity of 1.5 km s^{-1} corresponds to a Doppler shift of 0.003 nm. For such shift to be detectable by your instrument, it needs a resolution of about 10 times less (thus 0.03 nm), since we can measure the position of a line to about the precision of a tenth of a pixel. For example, in the red at around 650 nm, it needs to have a resolving power higher than:

$$R = \frac{650}{0.03} = 21,600$$

This is somewhat a rough order of magnitude. There are tools which vastly increase the sensitivity if we use a large spectral domain containing a large number of lines (cross correlation functions). But do not hope to see the revolution of a binary star (a few ten km s^{-1}) with a spectroscope of resolving power $R = 600$: you will not be able to separate the spectra of the two stars.

I can very briefly summarize the resolution needed for each kind of physical phenomenon:

- radial velocities (Doppler effect) are generally reserved to high resolution instruments, unless you consider very large velocities, as for galaxies or quasars;
- stars' temperatures are for low resolution, to have a large spectral domain and be able to measure a Planck profile;

– anything related to spectral line profiles requires high resolution;

– anything related to the object identification (research of its lines, spectral type) is suited to low resolution, since this allows for a larger spectral domain.

4.4 Start with Low Resolution

You might feel lost in front of all these possibilities. If you don't know where to start from, and if you can still chose your instrument, my advice is simple: start with low resolution. The observations are easier because they are shorter, and you will be able to quickly go over all the objects mentioned before: there is a lot to see and learn. It is also globally cheaper, because a "low-resolution" spectroscope is more compact than a "high-resolution" spectroscope (we will see why later).

For several reasons (the BeSS program once again!), it happened that the first spectroscope commercialized by the company Shelyak Instruments (the Lhires III) was a high-resolution instrument ($R = 18,000$). Looking back now, I see that the learning curve to jump in at high resolution is quite steep: you need a lot of abstraction to enjoy a spectral line profile when you have never seen a complete spectrum in the visible range.

4.5 Start with Different Spectral Types

In the same spirit, it seems to me obvious that your first observations should start with the different spectral types. It is really the heart of astronomy, and you will immediately see – on bright and easy to observe stars – the diversity of profiles offered by *Mother Nature*. You just need to observe a hot star (for example *Vega* – a favourite of spectroscopists) and then a cool red star (for example *Arcturus* or *Betelgeuse*) to understand that stars send us rich and easy to decode information.

4.6 Organize your Observation

I will keep on repeating this regularly: the first condition for the success of your observation is to prepare it. Decide ahead of time the object(s) you are going to observe. It is too sad to spend hours to set up the instrument, and once it is ready, ask what you could possibly observe. Conversely, if you have a clear and simple objective, it will guide all your actions during the night.

Chapter 5

Optical Principles
of a Spectroscope

After the brief review of the objects we can observe, I will now focus on the instrument that allows us to "see" their spectra: the spectroscope. I will begin by outlining a few basic rules useful in optics, and then I will describe the two main disperser elements available (prism and grating). I will also refresh a few principles of geometric optics needed to understand how the instrument functions. Finally, I will describe, building on this basis, the structure of a spectroscope and illustrate a real example. I will end this chapter describing the guiding stage and the light sources used to calibrate the spectroscope.

5.1 Reflection, Refraction, and Diffraction

All the optics we need fits in these three phenomena: reflection, refraction, and diffraction. The first two are the basis of geometric optics.

Reflection

It is the simplest phenomenon: a light ray reaching a reflecting surface at an angle θ with the direction perpendicular to the surface is reflected with the same angle. This is the first Snell-Descartes law (fig. 5.1).

This phenomenon is used by every mirror – the one in your bathroom as well as the one in your telescope. Reflection is exactly the same, independent of the wavelength of the light ray. (we say that mirrors are achromatic).

Refraction

When a light ray crosses the interface between two distinct materials (for example air and water, or air and glass), it is deviated. The amount of deviation depends on the refractive indices of the two materials (fig. 5.2). The refractive index is an intrinsic parameter of the material – usually

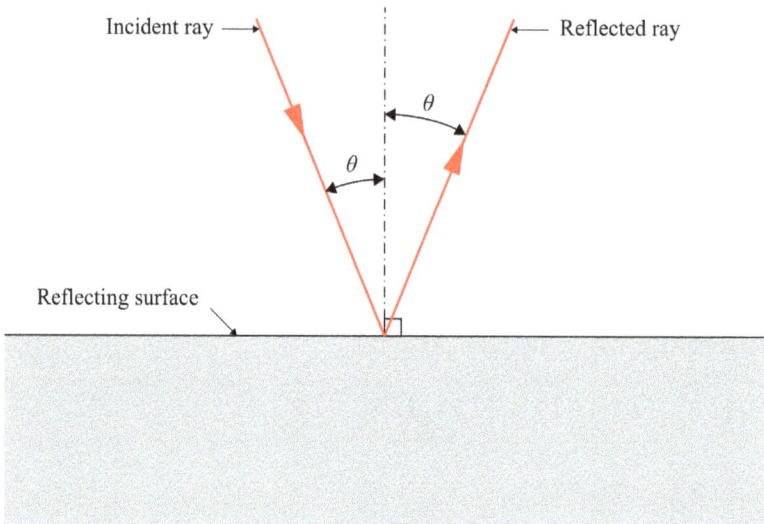

FIG. 5.1 – Reflection.

indicated as n, and its value is 1 for a vacuum and approximately 1 also for air.

Let θ_i be the incident angle with respect to the direction perpendicular to the interface, then the angle θ_r of the refracted light ray is given by:

$$n_i \sin \theta_i = n_r \sin \theta_r$$

where n_i and n_r are the refractive indexes of the material in which the incident and refracted ray propagate, respectively. This is the second Snell-Descartes law.

Actually, nature is slightly more complicated. When the incident ray hits the interface, only a portion of its energy crosses and becomes the refracted ray; the rest is reflected according to the first Snell-Descartes law. What fraction is refracted depends on the incidence angle and on the two materials.

Moreover, the refractive index of a medium is not exactly the same for all wavelengths. The good news is that this causes light dispersion in a prism, which enables us to form spectra. The bad news is that this causes the chromatic aberration so common in optics.

Refraction is the physical basis of the functioning of optical lenses (I will come back to them later).

Diffraction

Reflection and refraction are the most important optical phenomena *on large scales* – that is when the characteristic length of the instrument is larger than (say, 100 times) the wavelength of the incident light rays. We know that

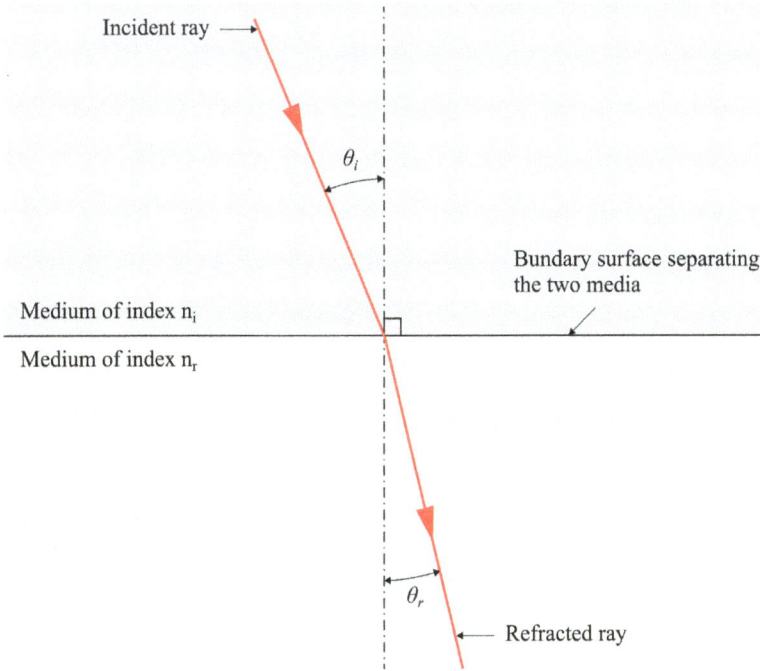

FIG. 5.2 – Refraction.

visible light, which is our concern, has wavelengths in the range from 380 nm to 1,000 nm. Thus, these two laws are sufficient for instruments larger than 1,000 nm × 100 = 100 μm = 0.1 mm. This is the domain of geometric optics.

On scales smaller than tenths of millimeters, the wave behavior of light takes over, and we observe the phenomenon of diffraction. The waves interfere with each other, and this causes surprising behaviors. For example, if we shine light from a lamp onto a small hole in a piece of paper, light exits the hole in all directions (one could expect it to exit along the direction of the lamp) (fig. 5.3).

If we have two holes close to each other, then we see interference fringes alternating light and dark regions (fig. 5.4).

A spectroscope combines the two scales: the lenses belong to the geometrial optics domain, while the diffraction grating is characterized by patterns of a few nm and uses the wave nature of light.

5.2 Prism and Grating

The aim of a spectroscope is to transform light from a source into its spectral components. It is a purely optical instrument, functioning according

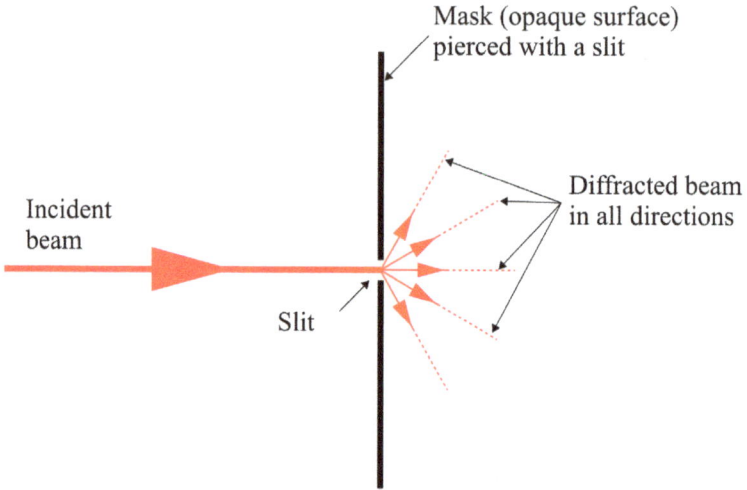

FIG. 5.3 – Diffraction from a hole.

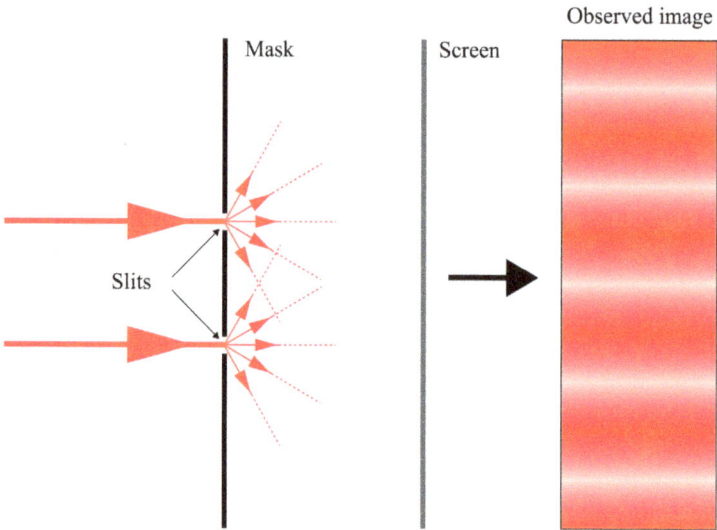

FIG. 5.4 – Diffraction pattern.

to very simple principles. In the heart of the instrument, we use a dispersion element to deviate the light depending on its wavelength. The vast majority of all spectroscopes use one of the two existing kinds of dispersion elements. The first kind is widely known: the prism. The second kind is a diffraction grating (fig. 5.5).

FIG. 5.5 – Photo of a prism (upper) and of a diffraction grating (lower).

In some instruments, there is a combination of the two elements, the so-called grism (fig. 5.6)

Prism

A prism is a block of glass with a triangular shape, with at least two polished faces. When a beam of monochromatic parallel light rays (i.e. of one single wavelength, a pure color) enters one of the faces, it exits deflected from the other side, and the light rays are still parallel to each other (fig. 5.7).

Generally, we use prisms at their minimum deflection, whose angle is defined by

$$\sin i = n \sin \left(\tfrac{A}{2} \right),$$

where i is the incidence angle of the beam on the first face, A is the apex angle of the prism, and n its refractive index. This index depends on the glass the prism is made of, together with the wavelength considered. The dependence on wavelength is useful: a red beam is deflected by a different angle than

FIG. 5.6 – Example of a grism, a combination of prism and grating.

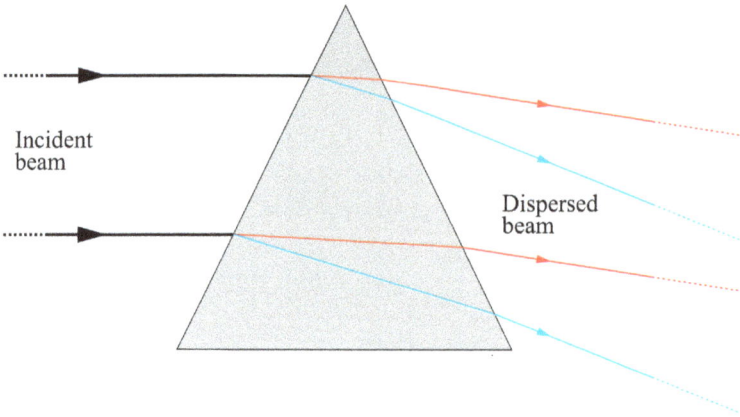

FIG. 5.7 – How a prism works.

a blue beam. And if an "ordinary" – i.e. non-monochromatic – light beam enters the prism, each color in it exits at a different angle. The differences in the angles of each color are typically small compared to the average deflection angle, but sufficient for our applications. Of course, we use prisms made of glasses with the strongest dispersion properties.

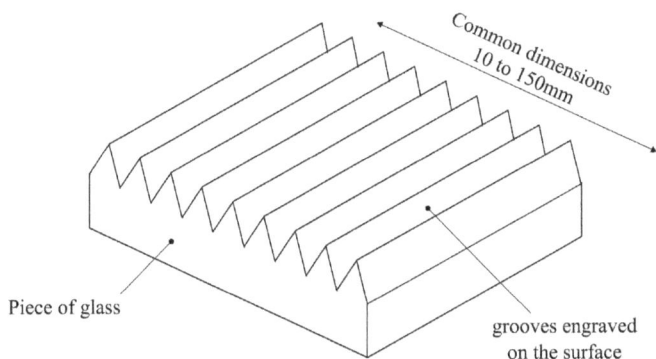

FIG. 5.8 – Scheme of a grating.

Diffraction Grating

A diffraction grating is a block of glass with a multitude of very tight grooves – gratings with 100 l.mm^{-1} to 3,600 l.mm^{-1} (lines per millimeter) are common. There are transmission gratings (which allow the light to pass through) and reflection gratings (which reflect the light, as mirrors). Both use the same principle, and only the structure of the instrument determines the need for one or the other. Figure 5.8 shows the structure of a reflection grating.

Diffraction gratings are nowadays very common optical components (the first were built and used in spectroscopy by Joseph Von Fraunhofer at the beginning of the nineteenth century), but they are nevertheless technologically amazing. They require etched lines separated by a few tens of nanometers with a precision of the order of nanometers. They are not well known by the general public, but they literally caused a scientific revolution, since they are the heart of most spectroscopes – not only in astronomy.

The distance between the etched lines can reach a few hundred nanometers, which means we are in the domain of diffraction. On this scale, each etched line can be considered as a slit from which light exits in all directions.

If we shine a beam of monochromatic light (i.e. of given wavelength) onto such a grating, the exiting light is as if it were produced by a multitude of very small light sources separated by the width of the lines. To simplify, let's consider two rays R_1 and R_2 falling onto two adjacent grooves of the grating separated by a distance d, with an incidence angle θ_i with respect the perpendicular to the grating. The rays are diffracted in all directions (fig. 5.9).

Now, let's consider one of these diffracted rays, for example the one exiting at an angle θ_r in the following figure (fig. 5.10).

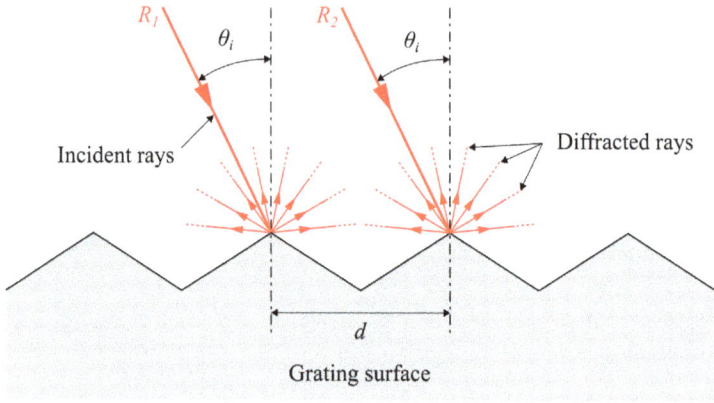

FIG. 5.9 – Diffraction of monochromatic light.

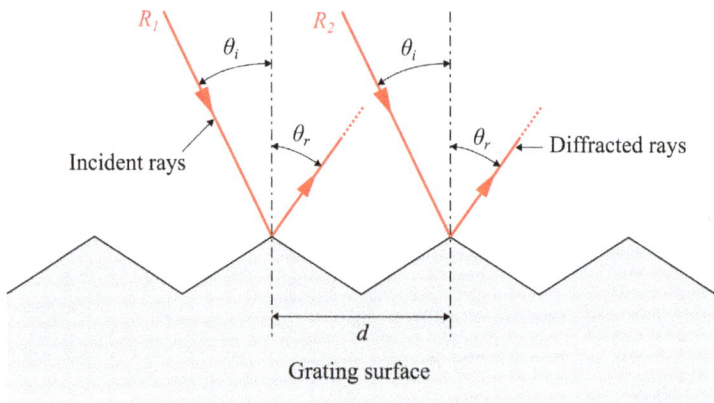

FIG. 5.10 – Schematic representation of a diffracted ray.

Let's look at the *optical path* (distance) traveled by each of these rays (fig. 5.11).

At the grating exit, the ray R_2 is lagging ΔR behind the ray R_1. This lag is the algebraic sum of the lags ΔR_i between the incoming rays, and the lag ΔR_d between the outgoing rays. In total, ΔR is given by the trigonometric relation:

$$\Delta R = d \cdot \sin \theta_r + d \cdot sin\theta_i \text{ (the angles have opposite signs in the figure below)}$$

(where d is the distance between the two grooves).

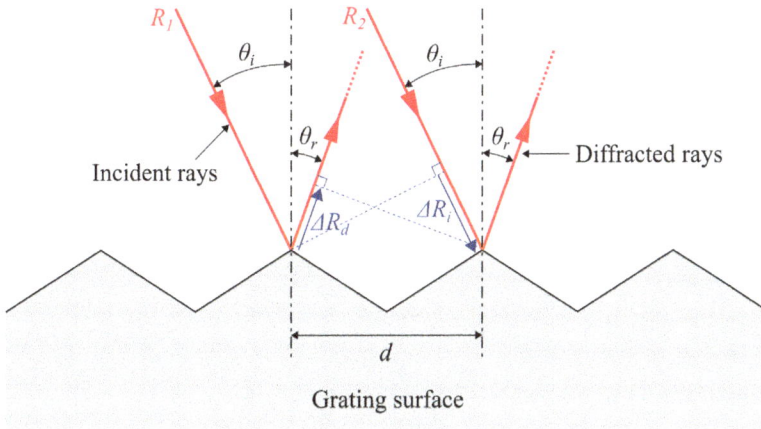

FIG. 5.11 – Optical path of the two beams.

Recall that the rays are electromagnetic waves, and the two waves can add or cancel each other depending on their phase. If the lag ΔR is exactly equal to a wavelength, then the two rays will add up. In other words: for any given wavelength λ and incidence angle θ_i, there is an exit angle θ_r for which the spatial lag ΔR is exactly equal to the wavelength λ. This happens when

$$\Delta R = \lambda,$$

that it $\sin\theta_r + \sin\theta_i = \lambda/d$

If now we look at the ensemble of the rays from each groove of the grating, all of them are superposed at that specific angle θ_r. For all the other angles, each groove of the grating gives a different lag, and since there are a large number of rays, on average they will cancel each other. Therefore, the wavelength λ exits only at angle θ_r. Similarly, another wavelength λ' exits only at a different angle θ_r'.

This is the trick of the grating: if we illuminate it with white light (containing all the wavelengths), the exiting light is dispersed – with an angle θ_r corresponding to each wavelength λ (fig. 5.12).

Diffraction Order

Rays coming from different grooves in the grating add together when the lag of each ray with respect to the neighboring is equal to the wavelength. The same thing happens if the lag is twice the wavelength, or k times it. This means there exist a second angle θ_{r2} for which all the rays are superposed, and actually, there will be a whole multitude of angles θ_{rk}. Each of these angles corresponds to a different *diffraction order* (fig. 5.13).

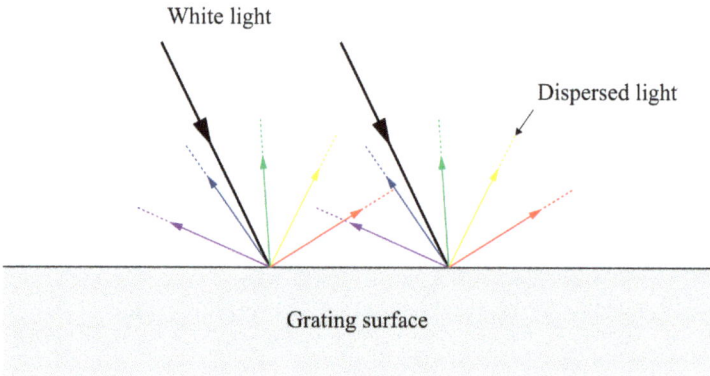

FIG. 5.12 – Dispersion of white light.

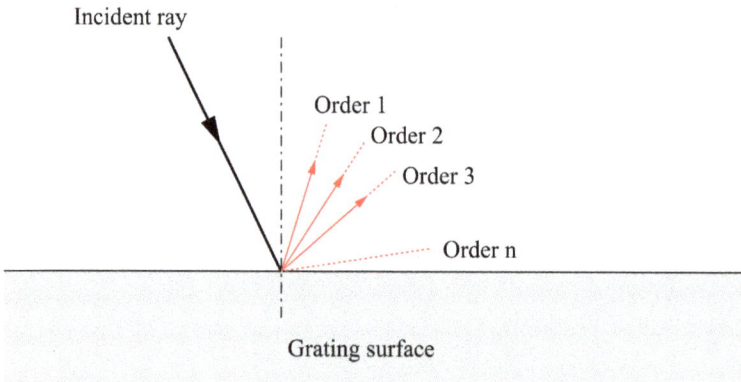

FIG. 5.13 – Several diffraction orders.

If we consider the diffraction order, the equation above becomes:

$$\sin \theta_r + \sin \theta_i = k\lambda/d$$

where k is an integer corresponding to the diffraction order.

d is the distance between two lines of the grating. Typically, it is more common to define a grating using the number of lines per millimeter n rather than using the distance between two lines. If d is expressed in millimeters, then $d = 1/n$.

Consequently, we can rewrite the equation for a grating in its most common form:

$$\sin \theta_r + \sin \theta_i = nk\lambda$$

where we have to take care of using wavelengths in millimeters. The notion of diffraction order has several important consequences for our application.

First of all, on the negative side, all the energy of the light conveyed toward the grating will not be focused on a particular diffraction order, but instead it will be spread on all the possible orders. Since most spectroscopes work only with one order (typically the first one), it is easy to foresee an efficiency problem – that prisms do not have. This is partially solved with a design trick (Blaze angle, see below), but is something to keep in mind.

Then, in some cases there can be a problem of orders overlapping. Take for example a spectroscope covering the entire visible domain (from 380 nm to 800 nm). An extremely blue ray (at 380 nm) has a diffraction angle of the second order equal to the first order diffraction angle of a near infrared ray at 760 nm. This means that, at this precise angle, we see a mixture of the blue spectrum at the second diffraction order and of the red spectrum at the first order.

This is a very serious problem which intrinsically limits the spectral domain accessible to a grating spectroscope. In practice, we can use an *order filter* on our spectroscope to block the red or the blue part depending on the part of the spectrum we are interested in; but we cannot see the whole spectrum at once.

On the other hand, this business of order superposition is exploited in a very particular, but rather complicated, kind of instrument: the échelle spectroscope.

A Numerical Example

Let's get out of the theory to have a look at a real case. Suppose we have a reflecting grating of 2,000 l mm^{-1}, and to shine light at an angle $\theta_i = 60$ degrees on it. What are the first order diffraction angles for an extreme blue beam ($\lambda_1 = 380$ nm), a green beam ($\lambda_1 = 550$ nm) and a red beam ($\lambda_1 = 650$ nm)?

We can use the grating equation:

$$\sin\theta_r + \sin\theta_i = nk\lambda$$

Since $\theta_i = 60$ degrees and we are looking for the first order ($k = 1$) diffraction of a grating of 2,400 l mm^{-1}, the equation becomes:

$$\sin\theta_r = (2{,}400 \times \lambda) - \sin(60 \text{ degrees})$$

Therefore, the diffraction angle is:

$$\theta_r = \arcsin\{(2{,}400 \times \lambda) - \sin(60 \text{ degree})\} \text{ (where } \lambda \text{ is in mm)}$$

For the blue light, we have $\lambda_1 = 380$ nm $= 380 \times 10^{-6}$ mm. Thus:

$$\theta_{r1} = \arcsin(2{,}400 \times 380 \times 10^{-6} - 0.866) = 2.64 \text{ degree}$$

Similarly, for the green $\lambda_2 = 550$ nm $= 550 \times 10^{-6}$ mm. Thus:

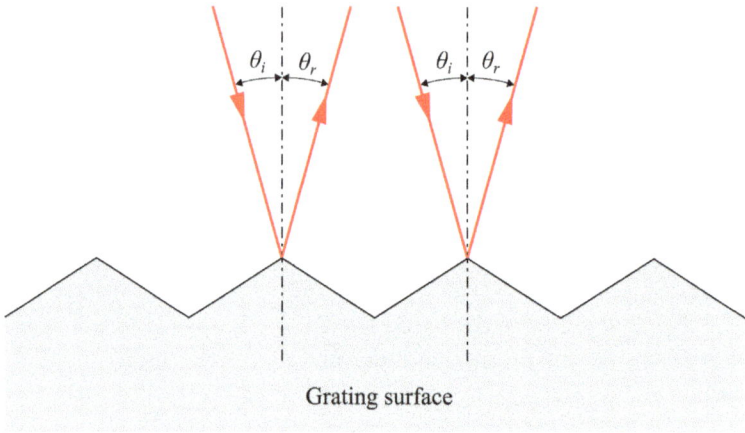

FIG. 5.14 – The zero diffraction order.

$$\theta_{r2} = \arcsin(2{,}400 \times 550 \times 10^{-6} - 0.866) = 27.0 \text{ degree}$$

Finally, for the red $\lambda_3 = 650$ nm $= 650 \times 10^{-6}$ mm. Thus:

$$\theta_{r3} = \arcsin(2{,}400 \times 650 \times 10^{-6} - 0.866) = 43.9 \text{ degree}$$

Therefore, a grating with $2{,}400\,\mathrm{l}\ \mathrm{mm}^{-1}$ disperses the visible domain over more than 40 degrees.

Zero Order and Negative Orders

There is a very particular diffraction order: the zero order (fig. 5.14). If the lag between a ray and its neighbor is zero, there will of course be a superposition of those. This happens when the diffraction angle θ_r is equal to the incidence angle θ_i – i.e. when the grating acts as a simple mirror.

When this happens, it is for all the wavelengths: this means that for the zero order there is no dispersion. In reality, with a grating spectroscope, you will always see a non-dispersed image of the source and, next to it, one or more diffraction orders (fig. 5.15).

I have defined the diffraction order using the lag between two rays, but a ray can *precede* another. In such cases, we have a negative order – effectively, the zero order is surrounded by positive and negative orders (fig. 5.16).

Blaze Angle

Grating makers not only are capable of creating gratings with an incredible line density: they can also shape these lines with a specific profile. In reality,

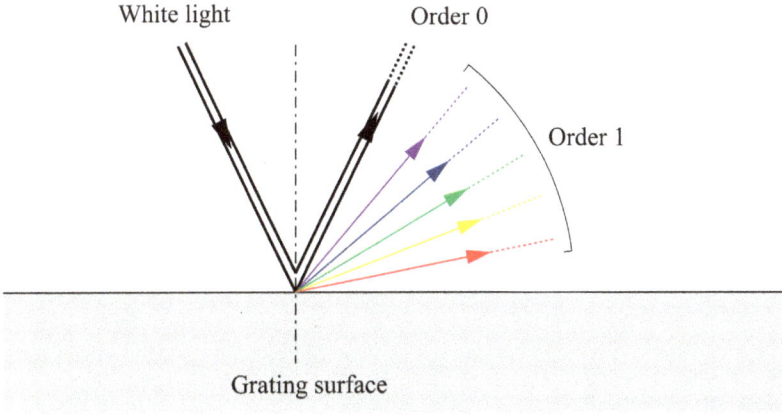

FIG. 5.15 – Zero order for white light.

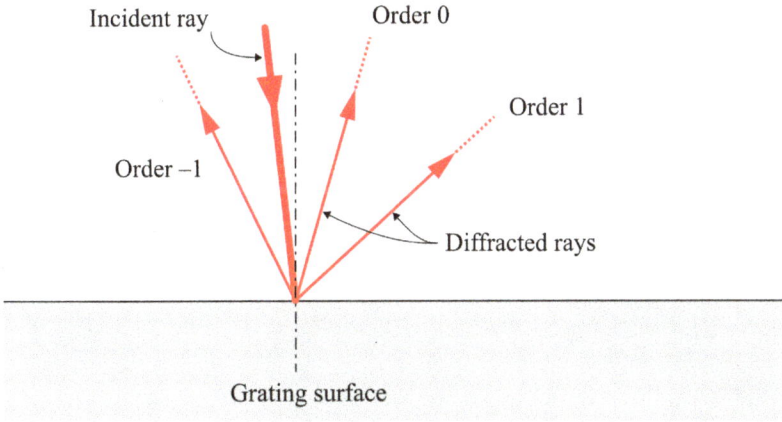

FIG. 5.16 – Negative diffraction order.

the lines in a grating are seldom symmetric, and more often they look like the scheme in figure 5.17.

I said that the phenomenon of diffraction leads to a spread of rays in all directions. This is certainly true, but not all the directions have the same intensity. By analogy with the light passing a small hole, there will always be more energy along the axis of the hole than in deviated directions.

We can exploit this property by choosing a specific line profile for the grating. It is easy to see that for the direction normal to the surface of the grating lines, each line behaves as a mirror, and in this direction the diffraction

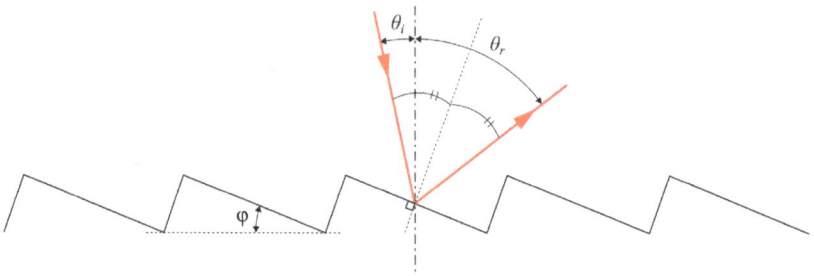

FIG. 5.17 – Blaze angle.

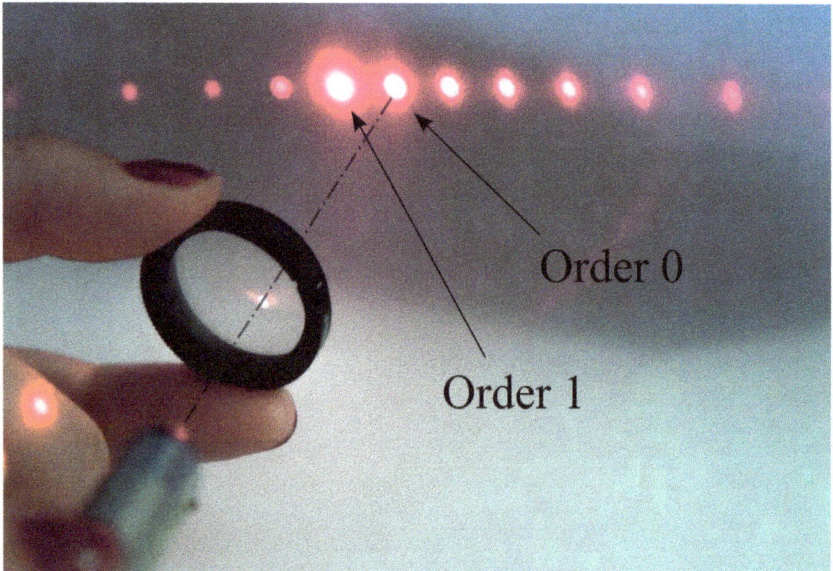

FIG. 5.18 – Several diffraction orders using a laser.

is more intense than in the others. This trick favors a certain diffraction angle: this is the so-called *blaze angle* (indicated with φ on figure 5.17).

The results are impressive: true, the energy of the light is spread over all orders, but choosing the blaze angle, we can strongly enhance one order (typically the first one). In the end, the efficiency of a grating in the selected order is much higher than if it were purely symmetrical.

As shown in the figure (fig. 5.18), if we point a laser at a transmission grating, we see all the different diffraction orders, and we can notice that one of them is clearly more luminous than the others.

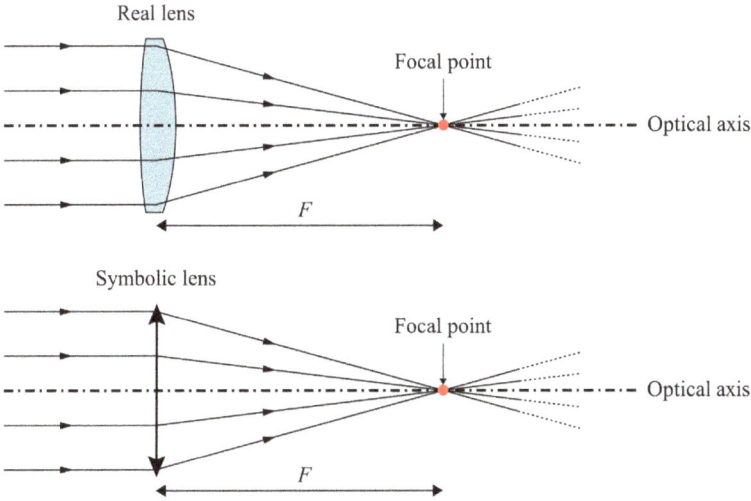

FIG. 5.19 – Thin lens.

5.3   Review of Geometric Optics

Together with the dispersion element, a spectroscope uses two optical components (collimator and objective), which can be approximated to the first order with a simple lens. A lens is a block of glass having at least one side polished with a spherical shape. This makes it able to converge rays coming from infinite in a point called focus. Often, a lens is represented by a vertical segment with arrows at its extremities (fig. 5.19).

The distance between the lens and its focus is called focal distance (indicated by F on the previous figure): it is the most important parameter of the lens. Another important parameter is its diameter D, which sets the maximum size of the beam that can fit into the lens. The ratio F/D between these two parameters is called focal ratio (or F-ratio): the smaller this ratio is, the more quickly light converges from the lens to the focus. For example, figure 5.20 represents two configurations with $F/D_1 = 5$ and $F/D_2 = 2$.

We can define some other useful elements, shown in figure 5.21, for the lens:

- the optical axis of the lens is its axis of symmetry;

- if a lens is symmetric: it actually has two foci (f and f') at the same distance from the lens;

- the focal plane is the plane perpendicular to the optical axis on which the focus lies;

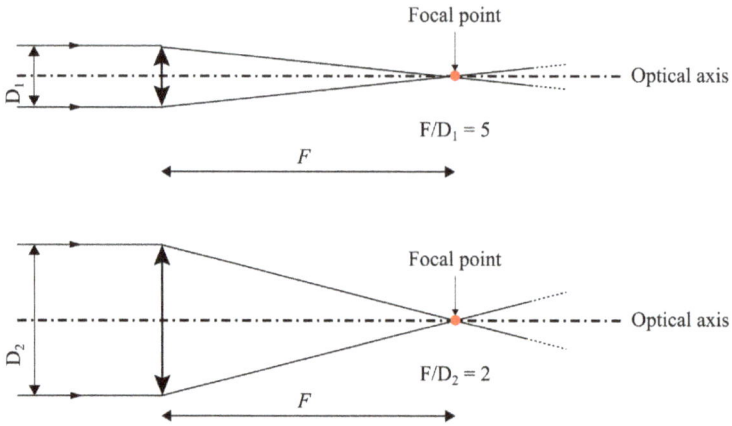

FIG. 5.20 – F, D, and F/D.

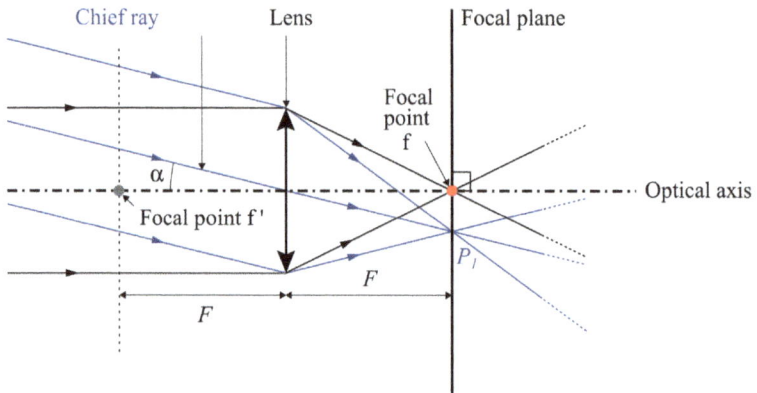

FIG. 5.21 – Scheme of a beam of light rays crossing a lens.

– we call the chief ray the one passing through the center of the lens;

– the chief ray is not deviated by the lens;

– if a beam of parallel rays enters the lens with an angle α with respect to the optical axis, it is focused on another point P_1 of the optical plane, easy to find by following the chief ray.

Therefore, we can consider the lens forming an image of an object at infinity, and if we put a detector in the focal plane (for example a CCD), we obtain an image of this scene. In spectroscopic use, the lenses are always used

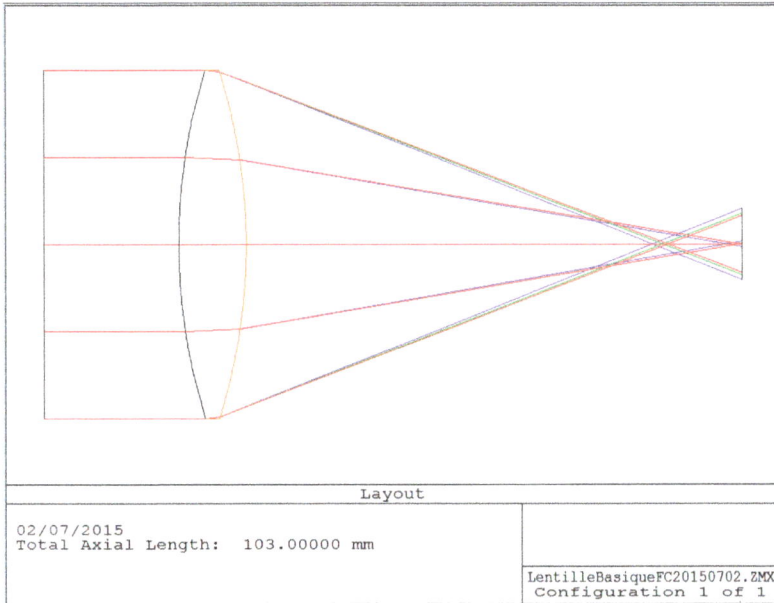

Layout

02/07/2015
Total Axial Length: 103.00000 mm

LentilleBasiqueFC20150702.ZMX
Configuration 1 of 1

FIG. 5.22 – Exact ray tracing of a beam passing through a lens (Chromatic aberration).

as follows: a beam of parallel rays is converged to the focus. A beam of parallel rays is a beam coming from infinity. Typically, we consider a source to be at infinity if its distance to the spectroscope is greater than one hundred times the focal distance. This is the case in astronomy, since we observe extremely distant objects.

In other contexts, for example photography, the lens can be used to form images of a scene relatively close – I will not spend time on this.

The whole scheme above can also be reversed. If we place a point-like light source at the focus of the lens, it will produce a beam of parallel rays. (We can say that the image of the source is projected to infinity).

The focal length of the lens depends on the refraction index of the glass used to make the lens, and on the radius of curvature of its faces.

Attention: I have so far described an ideal lens, which unfortunately does not exist in reality. We call that a paraxial lens. This expression (literally, "close to the axis") means that the closer we are to the optical axis, the more ideally the lens behaves. Real lenses are indeed polished with spherical surfaces – relatively easy to make – but if we make a precise calculation (or a measurement on an optical bench) of the projection of each ray in a parallel beam, we see that the convergence is not exactly on the focal plane for all of them. In reality, the situation is more like the one presented in figure 5.22. (This is produced using the Zemax calculation software).

The defects are increasingly more important when the parallel beam moves away from the optical axis. There are several kinds of defects (called optical aberrations): coma, astigmatism, field curvature, distortion, chromatic aberration, etc. The art of optics consists in correcting these defects using not only one lens, but many, each with complimentary characteristics. This is why, for example, photographic camera objectives have several lenses. There is no perfect optics – but for each specific need there is an optimal configuration. I am not an optician, but I have understood in the past few years that optics is a complex and noble topic, which requires a lot of technical expertise, experience, and creativity.

Always keep in mind that an optical setup is harder to correct if its F-ratio is smaller. This is the same as in photography: an objective with $F/5.6$ is relatively simple. A comparable objective (i.e. with the same focal length) but at $F/2.8$, is more complicated and far more expensive.

Luckily, when you use an optical instrument, you don't need to know all the details of the optical architecture – this is the designer's concern. But even if a real objective is made of several elements, it can always be considered, to the first order, as a single ideal lens, similar to the one described at the beginning of this paragraph.

However it is useful to know the main components of your instrument, to assure a good match with your telescope, and verify that it brings the performance you need for your observations.

5.4   Refracting and Reflecting Telescopes

To take astrophysical spectra you need a telescope, on which you mount your spectroscope. It is useful to know that the telescope can also be considered a paraxial lens, to the first order of approximation. As for the lens, the main parameters characterizing an astronomical instrument are its diameter and its focal length F_T. The diameter quantifies the amount of light that can be collected (the instrument is first of all a "light funnel"), and the focal length determines the size of the image of a star on the focal plane. When you use your instrument for visual observations, you put an eyepiece behind the telescope. The eyepiece (or ocular) itself can be approximated as a lens, with focal length F_O (fig. 5.23).

The magnification G of the instrument is then the ratio between the focal length of the telescope and of the ocular:

$$G = \frac{F_T}{F_O}$$

For imaging use, we substitute the ocular with the detector of a CCD camera in the focal plane of the instrument (fig. 5.24).

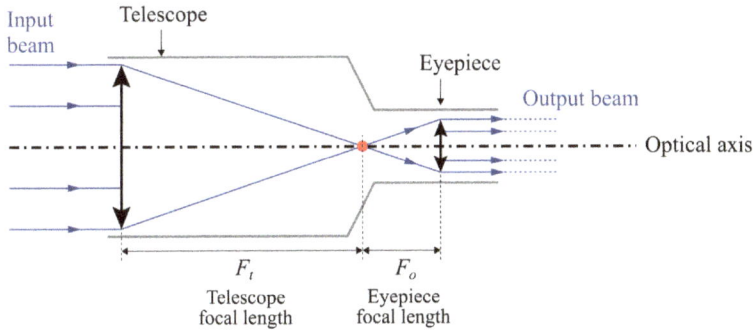

FIG. 5.23 – Simplified sketch of a telescope + ocular.

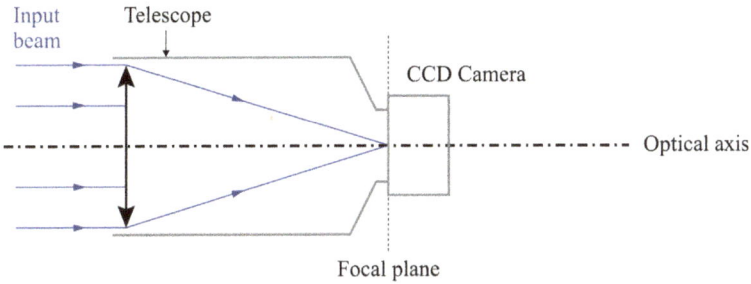

FIG. 5.24 – Simplified sketch of a telescope + receiver.

In the case of spectroscopy, we use the fact that the image of a star is almost entirely concentrated in one single point – the telescope focus – and use this point as our point-source (fig. 5.25).

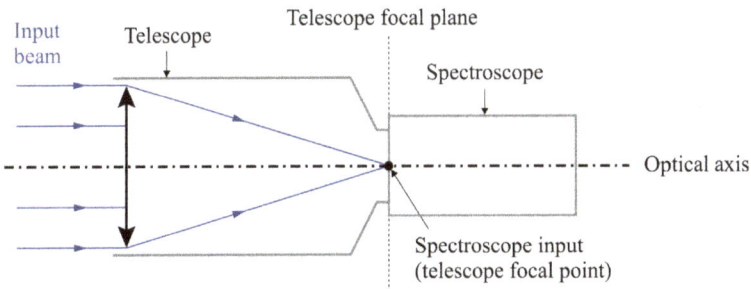

FIG. 5.25 – Simplified sketch of a telescope + spectroscope.

A telescope can be approximated as paraxial lenses. Once again, this approximation has its own limits. If they really were paraxial lenses, there would be no difference between telescopes. However, these differences do exist: they are caused by different qualities of the optical corrections. We expect a good telescope to yield a good image quality over the entire observed field: each star image – even those in the corners of the field – must be a point without chromatic aberration.

Chromatic aberration is an optical defect which results in a different focus for rays of different colors (along the same optical axis). In other words, the focal plane is not exactly the same for all wavelengths. Since this is color-dependent, it is easy to understand that it is an important issue in spectroscopy.

On the other hand, the image correction at the boundaries of the observed field is less relevant for our application: the object observed spectroscopically is generally on the optical axis – i.e. where the defects are smaller.

The degree of chromatic aberration is a criterion to consider when choosing an instrument to purchase. The fact that blue rays from the star come to a focus at a different position than the red rays can create a chromatic aberration also in the light entering the slit (for example by letting in more red light than blue light). This can have a significant impact on the final spectral profile, especially for short exposure times (on a long exposure time the effects are averaged and easier to compensate).

Generally, reflecting telescopes have less chromatic aberration than refracting ones, because they rely on mirrors (a mirror has no chromatic aberration). This is particularly true for Newtonian telescopes, which have no lenses, but instead have two mirrors, one parabolic and one plane (a Schmidt-Cassegrain has an initial corrector lens and then two reflecting mirrors). If you get the chance to use this type of telescope, do not hesitate.

This is not to say that we should ban the use of refractors in spectroscopy: they have other advantages. If you want to do so, you just need to use good quality refractors (they are common nowadays).

5.5   Architecture of a Spectroscope

After describing the elements necessary to build a spectroscope, I can now assemble them.

At the focus of the telescope we have a point-like image of the star. We want to transform this point into a beam of parallel light rays to have the optimal light to illuminate the dispersion element. We use a collimating lens (or collimator) for that. The beam then passes through or reflects off the dispersion element, and is dispersed into several parallel beams of different colors. These need to be focused again onto the detector to form the spectrum. We use a second lens, called an objective, for this. Each color of light converges onto a different point of the detector. Figure 5.26 shows the ensemble.

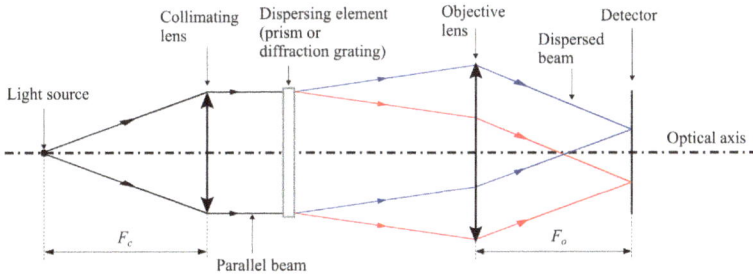

FIG. 5.26 – Architecture of a spectroscope.

This scheme can be designed in many different ways, but all these elements are found in every single spectroscope. You will often hear about collimators, prisms or gratings, and objectives.

For any given wavelength, this scheme illustrates the spectroscope's only function: to project an image of the source onto the detector – possibly with a magnification or reduction, depending on the ratio between the focal lengths of the collimator and the objective.

Let's consider a concrete example. Imagine a spectroscope with a collimator of focal length $F_c = 200$ mm, and an objective of focal length $F_o = 100$ mm. If the image of the source has size $d_s = 0.1$ mm, then the image projected on the detector has a size d_i

$$d_i = \frac{F_o}{F_c} \times d_s = \frac{100}{200} \times 0.1 = 0.05 \text{ mm}$$

Often, the possibility of playing with the magnification of the image is used during the spectroscope design to match the physical size of the spectrum to the dimension of the detector and its pixels.

In the scheme above, the collimator and the objective have an almost symmetrical role: both are used to "conjugate" (this is the optics technical term) the infinitely distant source on one hand and the focal plane on the other (or the other way around). However, in practice, they are often very different optical devices: the collimator works only on the optical axis (since the source is by definition on this axis), while the objective lens has to work across the focal plane to bring a parallel beam of light, for each wavelength, to focus on the detector. Along the optical axis, simple optics might suffice (for example an achromatic doublet), while the objective lens requires more sophisticated optics to compensate the chromatic aberrations. This is the reason why you might see photographic objectives (very good at compensating for chromatic aberrations) used in spectroscopes.

The Spectral Resolution

Now that we have a scheme of the spectroscope, we need to deal with the instrument resolution: how good is it in seeing details in the spectrum? We saw in chapter 3 that this is important to be able to observe physical phenomena. It is intuitive to see that the more the grating disperses light, the finer the details that can be seen. But this is not enough: also the size of the light source impacts directly on the resolution. If we do not pay attention, the apparent size of the star's image can change in time, e.g. :

- if the telescope is not optimally focused;
- if the atmospheric turbulence is significant;
- if the telescope tracking is not optimal;
- if there is wind, etc.

These various phenomena can significantly alter the resolution of the instrument – it is always a pity to not be able to exploit the nominal performances.

I will come back on how to measure the resolution of the spectroscope in chapter 6, but be aware that problems are so common in real observations that most spectroscopes have a slit at the focus to compensate.

Slit as a Point-Like Source

The purpose of the slit is to isolate a thin strip of light in the observed field, and only this particular strip is then dispersed by the spectroscope. During the observation, we make sure that the target star is well aligned to the slit. Since this slit has a fixed position in the spectroscope, it mitigates the problems mentioned above (focusing, wind, etc.). More specifically, when these problems arise, they result in a temporary shift of the star image out of the slit, and consequently a loss in intensity, but not in resolution of the spectrum. In other words, thanks to the slit, the resolution becomes an intrinsic parameter of the instrument, and does not depend anymore on the observing conditions.

Moreover, the slit at the focus brings some important advantages:

- It allows one to isolate the observed star. It can happen that stars close to each other in the field of view are perfectly aligned, which leads to a superposition of their spectra. It is then impossible to distinguish them – but the slit allows us to block the star light we are not interested in and get rid of its superimposed spectrum.

- Similarly, the slit allows one to isolate the sky background from the star. The sky background is never perfectly dark, because of light pollution. When we do slit-less observations, each point in the field produces its

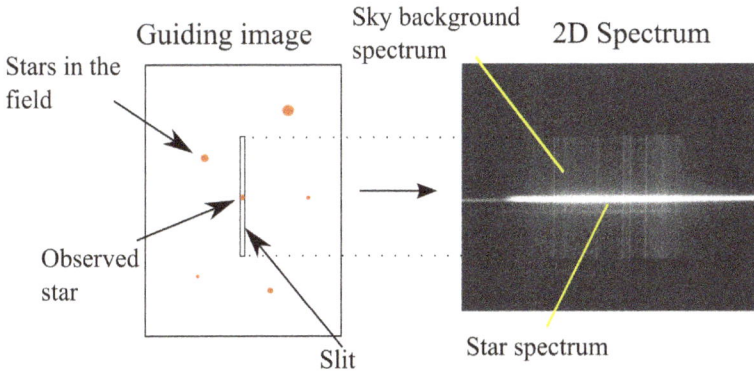

FIG. 5.27 – Schematic representation of a spectrum taken using a slit (with the sky background).

own spectrum on the detector. Thus, the background spectrum adds to the star spectrum, and it is impossible to separate them. The slit not only allows one to remove the background spectrum of the sky, but it also allows us to record it simultaneously while we take the stellar spectrum (fig. 5.27).

– It also allows us to take spectra of extended objects. I have only mentioned stars, which are point-like objects on the sky, but this is not true for all astronomical objects: comets, nebulae, galaxies...these are all extended sources. Without the slit, all the spectra from different points of the source are necessarily overlapped. With the slit, we can isolate a thin region of the source and take spectra at the nominal resolution of the instrument. Note that, in this case, the spectrum is now a large band representing parts of the extended object, and not a single line representing a point source anymore.

– The calibration of the spectrum is greatly simplified. The slit not only allows us to take spectra of the calibration lamp (neon, argon, etc. lamps – which are extended sources), but it also allows us to place the spectrum exactly at the same position on the receiver, ensuring that each wavelength hits the same spot on the detector during the calibration and during the observation.

Because of all these reasons, any high performance spectroscope has a slit in the focal plane, except very particular cases. Finally, figure 5.28 shows the conceptual scheme of a spectroscope.

The slit has a lot of advantages, but it also has a big inconvenience: it masks the telescope field. This has many consequences: when pointing the

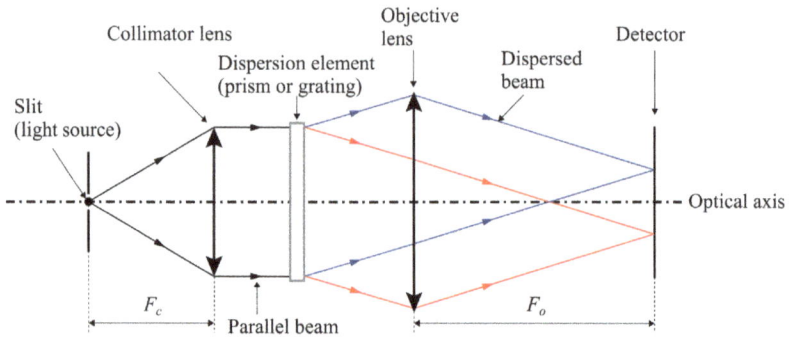

FIG. 5.28 – Architecture of a slit spectroscope.

telescope to a star, you work blindly. Imagine you had opaque glasses with a few micrometer wide hole in the center: you would have a hard time seeing anything through it. If you point your telescope to a star, and its image does not fall within the slit (which is very likely), you don't get any spectra and have no idea in what direction you should look for your star. To address this problem, we need to develop a complementary guiding system (see section 5.8).

5.6   A Real Example: Alpy 600

The conceptual scheme I presented in the previous pages is theoretical – any real instrument has peculiarities that deviate from this scheme. When designing the instrument, we of course consider the theory of the optics that goes into it, but we also have many other constraints, such as its size and weight, rigidity, the technical feasibility, the assembly, its price, etc.

Specifically, look at detail for the Alpy 600 and compare it to the conceptual scheme.

The Alpy 600 is one of the spectroscopes sold by Shelyak Instruments. It is an affordable, modular and compact instrument for low resolution. I focus here on the basic module, i.e. the spectroscope itself (the other main modules are for the guiding and calibration) (fig. 5.29).

In front of the spectroscope, you can see the slit – in reality the Alpy 600 has 6 different slits, which can be interchanged by turning the metallic plate. The slit "in use" is the one between the two screws. In figure 5.30, this is the 25 μm slit.

The cross section of the module fig. 5.31 shows the optical components.

The architecture is very close to the conceptual scheme described above: the collimator and the objective are simple achromatic doublets specifically designed for this application. The dispersion element is a *grism*, i.e. the combination of a diffraction grating and a prism. The grating is a transmission

FIG. 5.29 – The basic module of the Alpy 600.

grating, and it is made on the inclined face of the prism. Why a grism? Because a prism alone or a grating alone have the disadvantage of deviating the light beam by a large angle. Combining the two elements, the beam gets dispersed *and* deviated by the grating, and then the prism compensates the overall deviation to realign the beam to the optical axis of the instrument. Of course, the prism too has an effect on the spectral dispersion, but very small compared to the grating. Therefore, in the end, the spectrum is well dispersed but barely deviated. This allows to build the instrument around its optical axis, which is easier – the Alpy 600 contains primarily only elements with axial symmetry.

5.7 Another Example: Lhires III

The Lhires III is another spectroscope designed and commercialized by Shelyak Instruments – historically it was the first one designed by the company (fig. 5.32 and 5.33). It is a high-resolution instrument ($R = 18,000$).

FIG. 5.30 – The slit of the Alpy 600.

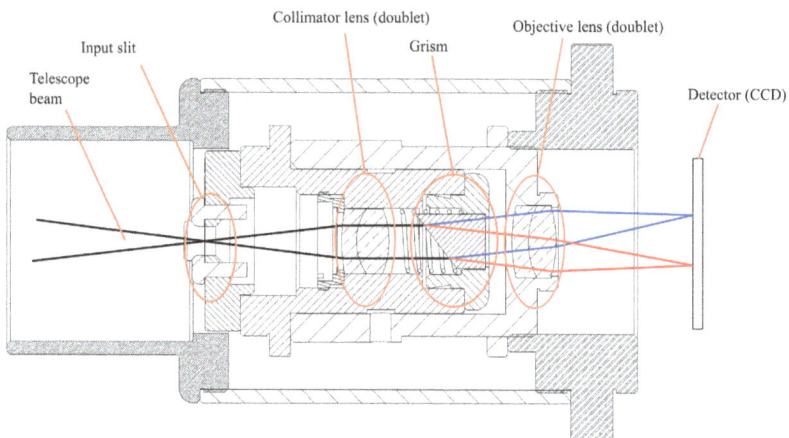

FIG. 5.31 – Cross section view of the Alpy 600.

The conceptual scheme is the same (slit - collimator - grating - objective - detector), but its technical design is quite different, since it has a *Littrow* architecture. The grating this time is reflecting, and the two lenses (collimator and objective) are merged into one single component: the same lens – an achromatic doublet – is crossed twice by the beam (on its way to the grating and after the reflection from it). This means saving money and weight in the optics, and provides a compact instrument with a folded architecture.

For this instrument we can use an achromatic doublet for the objective, since it is a high-resolution instrument: the spectral domain is narrow (of the order of hundreds of angstroms), and the chromatic aberration problems are very limited.

Of course, the optical design presented above is not on a single plane: if it were, the light dispersed by the grating would be sent back to the stars. A small tilt in the reflecting grating gives an angle between the incoming and the dispersed beam. The dispersed beam thus misses by a few millimiters the incoming beam. The effect of this small deviation on the quality of the spectrum is small, but the image of the slit when taking spectra of extended objects is noticeably bent – see for example the emission lines of neon (fig. 5.34). The bending has to be considered when reducing the data, otherwise the resolution is strongly downgraded.

However, this architecture has some disadvantages. The fact that we use only one optical component for collimator and objective imposes a magnification of 1:1 in between them. Thus, we lose the possibility of adapting the size

FIG. 5.32 – The Lhires III spectroscope.

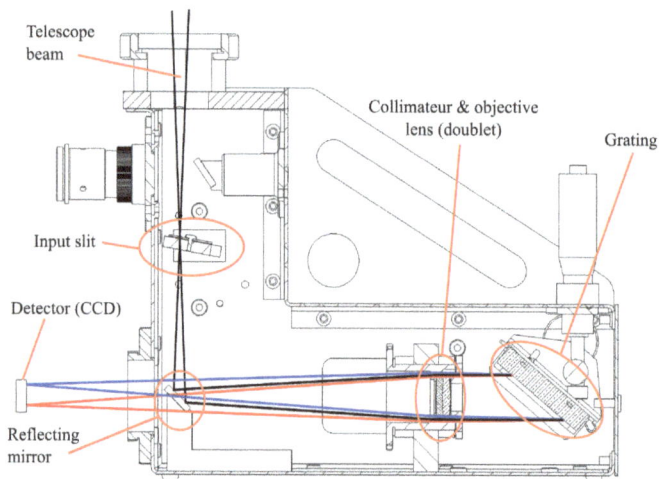

FIG. 5.33 – Cross section view of the Lhires III.

FIG. 5.34 – The emission lines are bent.

of the dispersed beam to the size of the receiver. In the Lhires III case, the situation is favorable: for a telescope of focal length 2 m (e.g. for a diameter of 200 mm at $F/10$), the size of the image of a star is of the order of 20 μm, i.e. a few pixels on a modern receiver. The situation would be different with a bigger telescope.

Another complication of the Littrow architecture is that the doublet has to be focused on the slit *and* the detector. In a classical architecture, each lens is focused separately. This imposes a very precise positioning of the detector with respect to the slit.

5.8 Guiding Stage

Above, I have explained that the slit offers many advantages for the spectroscope performances, but with the drawback of masking the telescope field of view. Most telescopes nowadays have an automatic pointing system (GOTO system) – it is almost a necessity when doing spectroscopy. Often, beginners think it is sufficient to point at the coordinates of the target star and trust the mount. Unfortunately, this is not how it works: the precision of pointing required is well beyond the capabilities of this type of mount. Picture the difficulty of your task: you need to place a star in a slit gap a few micrometers wide placed at the end of a focal length of several meters. Many phenomena create uncertainties of the same order of magnitude as the precision required. Among them:

- the rotation of the sky;

- the precision of your polar alignment;

- mechanical deformations of the telescope;

- atmospheric refraction (i.e. the prism-like effect of the atmosphere, which is stronger closer to the horizon)

Theoretically speaking, it is not fundamentally impossible: all these phenomena are deterministic and foreseeable. However, in real life, I have never seen – not even in professional observatories – an installation capable of pointing to a star right on the micrometer-wide slit relying only on the pointing capabilities of the telescope mount.

Regularly, other beginners think it is sufficient to place a guiding telescope aligned parallel to the telescope, as is common in astrophotography auto guiding. Once again, most of the people I know who tried this, gave up (some exceptions exist when targeting a bright star at low resolution). The differential flexure of the mount and the telescope are enough to put the star a few pixel off the slit, and we are once again blind. This technique works well in imaging because we look at the whole field of view and with relative motions, we do not need the absolute position of the star, as we do in spectroscopy.

The solution is to place a guiding system on the spectroscope. There are several kinds, but the basic principle is always the same: we get an image of the

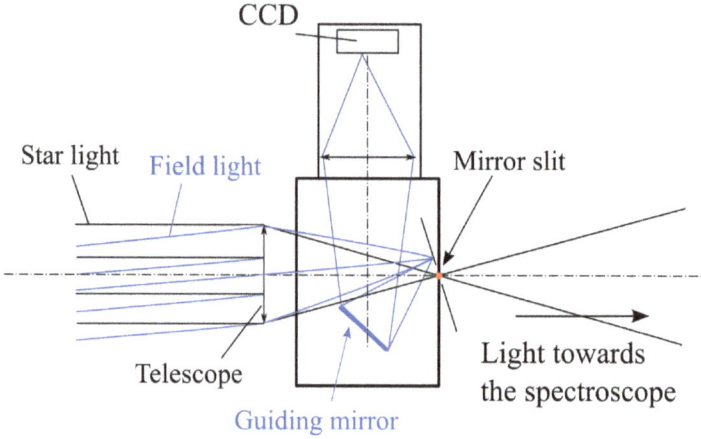

FIG. 5.35 – Conceptual scheme of a guiding system.

focal plane of the telescope before the light enters the slit of the spectroscope. In this image, you can see the whole telescope field of view, in addition to the slit.

Of course, we cannot put a camera in front of the slit: it would obscure the beam coming from the telescope. The trick is to observe the focal plane from a tilted position. Figure 5.35 illustrates the principle of such a guiding system.

The slit is located in a mirror slightly tilted. Thus, all the light (the image on the focal plane) not going through the slit is reflected off the mirror surface at this angle. On the side, we place a second mirror which reflects the light from the focal plane perpendicularly to the optical axis. A series of lenses "brings the image" to the guiding camera. The guiding image shows the telescope field of view and the slit, making it easy to find the position of the slit and put the image of the star in it during the observations: we are not blind anymore.

There are slightly simpler systems relying on a beam splitter: a fraction of the light coming from the telescope is re-directed to the guiding camera. These guiding systems are less handy, because we don't see the position of the slit. Moreover, only a portion of the incoming light is used for the guiding and another portion is used for the observation. With a mirror-slit, the whole flux from the star is used in the spectroscope and all the rest is sent to the guiding camera.

With a mirror-slit, we see exactly what happens at the entrance of the spectroscope: when a star's light enters the slit, it disappears from the guiding image – this is even a very good criterion to judge your setup. In practice, it never really disappears: there is always some leftover light that can be used to trace the exact position.

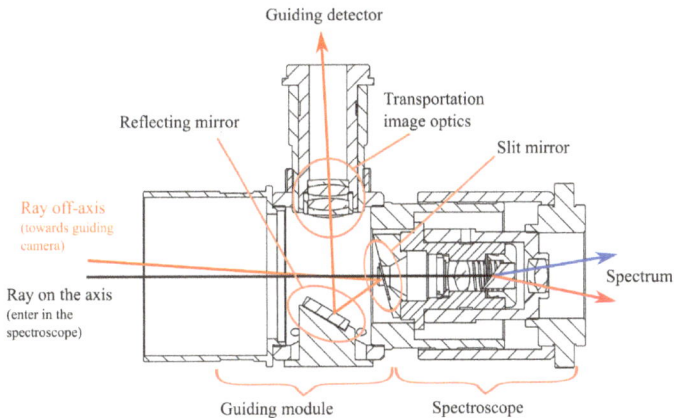

FIG. 5.36 – Module "Alpy guiding" (installed on the spectroscope Alpy 600 in the right panel).

The Alpy system has an optional guiding module ("Alpy Guiding Module"), which uses exactly this principle (fig. 5.36).

5.9 Calibration Light

The theory of optics allows one to find the dispersion and the spectral range of the instrument, but in real observations it is impossible to say beforehand what is the wavelength of the light going to each specific pixel of the detector. The dispersions during the production of the optical and mechanical parts, and even the deformation of the spectroscope itself in some cases, are much larger than the precision required.

There is a very effective solution to this problem: complete your observation by acquiring the spectrum of a known source, with sufficiently peculiar characteristics that you can safely recognize them. By chance, these sources exist: they are calibration lamps, with a specific gas – neon, argon, hydrogen, xenon, thorium, etc.

Analyzing the calibration spectrum we can later reconstruct the dispersion law and assign to each pixel a given wavelength.

To make the real observations more practical, most commercial spectroscopes offer, at least optionally, a calibration system. The basic principle is simply to illuminate the spectroscope slit with a calibration lamp.

In addition to the calibration light, this device also offers the possibility of taking a *flat* spectrum. It illuminates the slit with white light (from an tungsten lamp). It only has a Blackbody spectrum, without any lines. This Blackbody spectrum enables the compensation of the instrumental response when reducing the data. The optical principle is exactly the same, only the light source changes.

The calibration system is not strictly necessary when you start in spectroscopy. One could place a calibration (or flat) lamp in front of the telescope – it is less practical, but the result is almost the same. If you observe regularly, you will easily appreciate the integrated calibration system.

The Alpy system offers an optional module for calibration, which includes a neon-argon lamp, a white-lamp (tungsten), and a rotating screen that can cover the slit. This screen reflects the light from the calibration module into the slit, and also blocks the light from the stars during the calibrations.

For the sake of completeness on the question of the calibration light, let me add that in astronomy we have another light source with well known spectal lines: the stars themselves! They even have the advantage of having lines in the far blue (Balmer lines), where typical lamps emit almost nothing (see figure 5.37 for an example of hot star). Thus, in some cases, we can use the light from the stars – or possibly of the Sun – to calibrate.

When we observe a star from Earth, its light crosses the atmosphere. The atmosphere has its own spectral signature – mainly because of water molecules – which we also exploit quite often. In particular, the H_α line is surrounded by several atmospheric lines, the so-called *telluric* lines. They allow to quickly verify (and possibly correct) the calibration in a high-resolution spectrum (see figure 5.38 for an example with the spectrum of Vega). This trick is important because, since these lines are from the Earth's atmosphere, they are not affected by Doppler shifts!

21 Lyn - Apr 18th, 2015 - F. Cochard - Alpy 600 C8 Atik314L+

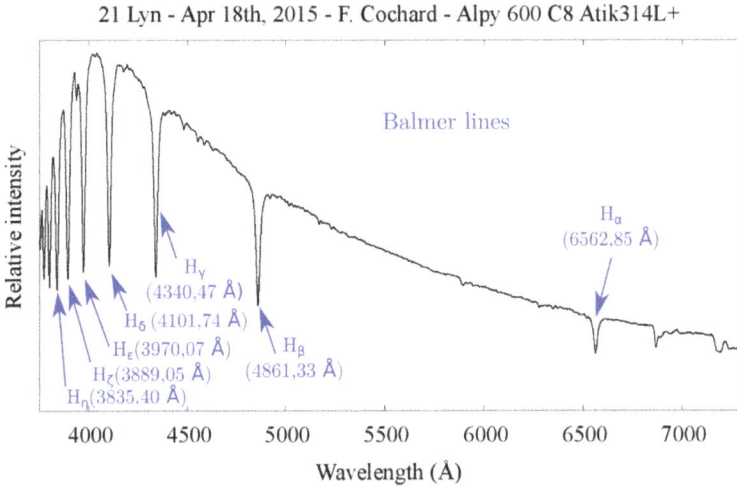

FIG. 5.37 – Balmer lines in the spectrum of a hot star.

Vega - 15-03-2014 - C. Buil - C11 Vhires (R=48000) - Castanet

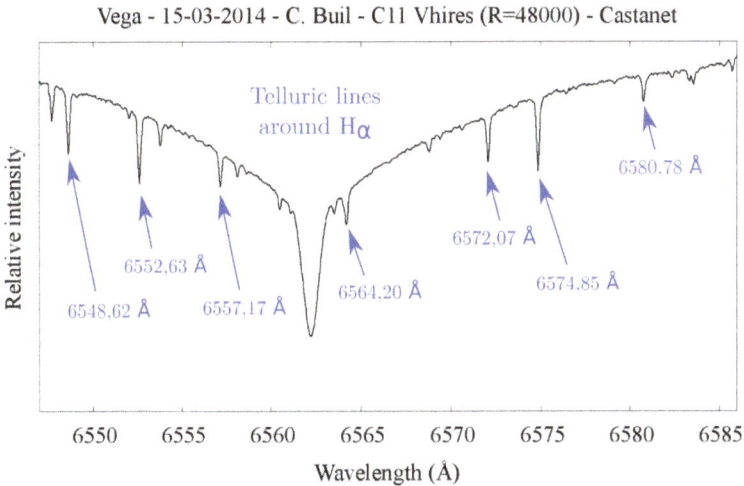

FIG. 5.38 – Telluric lines around H_α in the spectrum of Vega.

5.10 Échelle Spectroscopes

Let me digress, to describe briefly another spectroscope quite different in architecture, but also used in the amateur community: the "échelle" spectroscope.

Depending on the nature of the observation, we need to find a compromise between the dispersion and the spectral range, just because of the finite size of

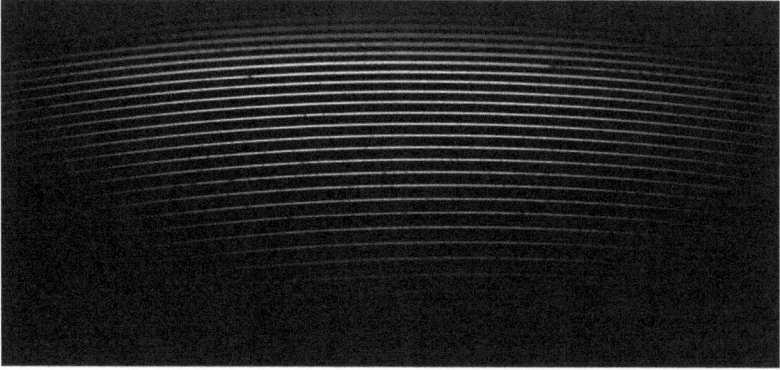

FIG. 5.39 – Raw image of an échelle spectrum of Venus (Solar light).

the detectors. Of course, there are situations where you need a high resolution (and therefore dispersion) and a large spectral range. The échelle spectroscope allows for this in a very clever way. The raw image from such an instrument looks like figure 5.39.

Each line is a portion of the spectrum, and they read like the lines of a book: the end of the first line connects to the beginning of the second, and so on. It is intuitive that using a larger surface of the detectors (instead of a single line of a few pixels wide) allows you to store a larger quantity of information.

The basic principle is the same as for an ordinary spectroscope: it uses a reflection grating for the dispersion. But instead of using the first order of diffraction, an échelle grating is used at very high orders (see section 5.2) – for example between order 30 and 50. Each line of the image above corresponds to one order.

This might look like magic, but there is however a size issue: these orders are superimposed when leaving the échelle grating, and we need to separate them later to see them individually. This separation is done using a second dispersion element – either a prism, or a grating oriented perpendicularly to the first one. Thus, we have a first dispersion to separate the light into different orders and then a second one to separate the orders. This creates an image somewhat complex to treat (the orders are curved), but all the information is there.

5.11 Fibre Optic

Another digression, on a topic slightly outside the framework of this book, but which is of concern for amateurs nonetheless. This topic is the use of optical fibres to transmit the light from the telescope to the spectroscope.

Optical fibres appeared in professional astronomy in the 1980s, together with very high-resolution spectroscopes.

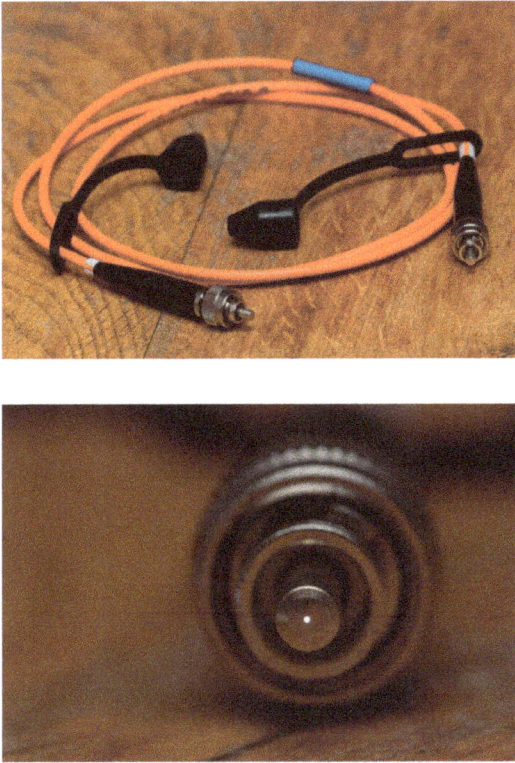

FIG. 5.40 – Fibre optic illuminated end.

An optical fibre is a "glass tube" that allows one to guide the light for a long distance with very little loss. An optical fibre looks like an ordinary electric cable, but glass fibers are substituted for the copper core (fig. 5.40).

The use of optical fibres allows us to physically separate the spectroscope and the telescope. This has important consequences:

– Since the spectroscope is not mounted on the telescope anymore, it can be very large: this is what contributes to the very high resolution;

– the instrument does not suffer from any differential flexure when the telescope moves. This simplifies the use, and improves precision, especially for velocity measurements.

Nevertheless, as always in optics, these important advantages have relevant drawbacks:

– The instrument is more complex, since you need a light collector system behind the telescope (the so-called guiding unit);

   – the fibres bring in specific constraints for the design of the instrument. For example, the beam needs to be fast (small F-ratio, say $F/3$) to work in good conditions;

   – the fibres do not have perfect efficiency: there are relevant leaks, especially at its ends.

In general, fibres optics allowed for the design of completely new instruments, that are high performance in many aspects. But they are not the cure-all, and paradoxically, the advantage of most of the amateur instruments is that they are compact and can be mounted directly on the telescope, for an intrinsically better performance. On the other hand, most professional spectroscopes use optical fibers, because of the size of the instruments.

Chapter 6

Main Parameters
of a Spectroscope

In this chapter, I will leave behind the theoretical aspects, and dive into practical considerations: it is time to do it. A spectroscope can be described with a certain number of key parameters that I am going to describe. These parameters are essential to adapt the spectroscope to the rest of your equipment (telescope, CCD), and to your observational targets.

One thing needs to be said straight away: *there is no ideal spectroscope!* Even if you invest a fortune in it, there will always be compromises to make depending on your own constraints. Take the analogy with telescopes: there is no instrument that allows you to do deep sky and planetary observations without giving up something. In deep sky observations, one aims for a large field without defects (optical aberrations, vignetting), while for planetary observations one would like the maximum magnification. These needs are contradictory, but many approaches are possible: either you go for a "universal instrument", or you invest in multiple instruments, each one optimized for different observations. The former approach minimizes the cost, sacrificing some quality in all the applications, the latter is obviously more expensive. Life is made of choices...it is what makes it so complex, but also so beautiful.

The two main parameters to consider are the resolution and the focal ratio. From these, the other parameters (sampling, slit size, spectral domain) can be derived, at least partially.

6.1 Resolution and Resolving Power

I have already mentioned the resolution several times: it is certainly the most important parameter of a spectroscope. Resolution defines the capacity of the instrument to observe details in the spectrum. More precisely, it is the dimension – in units of the wavelength (nm) – of the finest detail discernible in the spectrum.

In practice, it is more common to refer to the *resolving power* instead of the resolution itself. The resolving power R is defined as

$$R = \frac{\lambda}{\Delta\lambda} \tag{6.1}$$

where λ is the considered wavelength, and $\Delta\lambda$ is the smallest visible detail in the spectrum. Since both these quantities are length (expressed typically in nm), the resolving power R is a dimensionless number.

Let me give you some orders of magnitude. The lowest resolving power in amateur instruments are about $R = 100$ (very low resolution), and the highest are about $R = 20,000$ (high resolution). We call low resolution anything between $R = 500$ and $R = 1,000$ and medium resolution for $R = 1,000$ to $R = 5,000$.

If R is measured in the red (as it is often) – say at 650 nm, close to H_α – then a resolving power of $R = 100$ allows one to see details of $650/100 = 6.5$ nm. If you apply this to a measurement of radial velocity, you can measure the position of a line with roughly 10 times more accuracy, so 0.65 nm; which correspond to a radial velocity of 325 km s^{-1} (50 km s^{-1} correspond to 0.1 nm, see Section 2.4 about the Doppler effect).

Similarly, a resolution of $R = 20,000$ allows one to see details of $650/20,000 \simeq 0.0325$ nm, i.e. Doppler velocities of the order of 1.6 km s^{-1} (recall that the Earth is orbiting around the Sun with a velocity of 30 km s^{-1}).

To measure the resolution, we take the spectrum of a calibration lamp. The lamp shows emission lines at specific wavelengths, known a priori and well separated from each other. When describing the optical principles of the spectroscope, we saw that the spectrum is a series of images of the slit. In the spectrum of a calibration lamp, this phenomena is very easy to see (fig. 6.1).

If the instrument is well designed and set up, each line will end up only on a few pixels of the detector, forming a bell-shaped profile (Gaussian profile) (fig. 6.2).

We define the *full width half maximum* (FWHM) as a measurement of the width of the line (fig. 6.3).

If the instrument were *perfect*, the line would be infinitely thin. But perfection does not exist in this world, and the line is necessarily broadened by the optics of the instrument. The main reason for this broadening is the slit: it has a non-zero width (we need some light to pass through it).

Therefore, the FWHM is an indirect measurement of the smallest detail visible on the spectrum, which corresponds to the $\Delta\lambda$ in equation 6.1.

De facto, when a spectroscope maker announces a value of the resolution, it is determined for a given size of the slit. We will see later than we can play with the size of the slit, to adapt it to the telescope; but nevertheless there are limits and compromises to make.

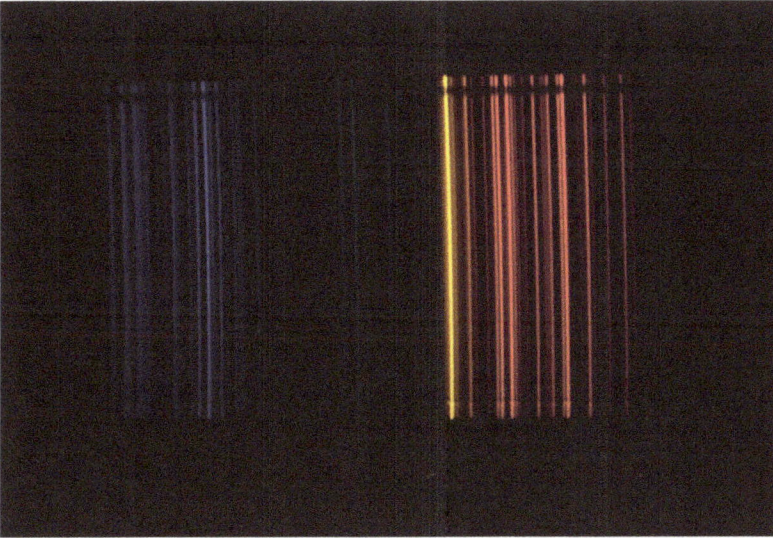

FIG. 6.1 – Spectrum of a calibration lamp.

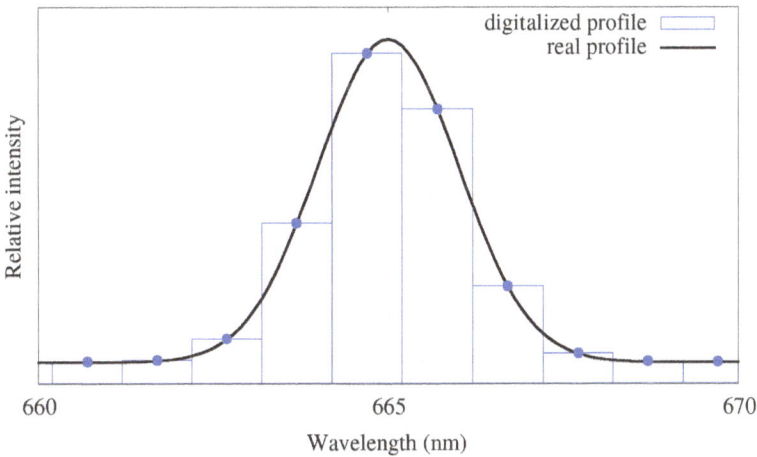

FIG. 6.2 – Projection of a line profile on the CCD pixels.

6.2 Focal ratio (F-ratio)

The focal ratio is the second essential parameter of a spectroscope. We can easily think of the telescope focal ratio as if it were a camera objective. The larger the telescope focal ratio, the more light it collects, and thus the shorter the exposure time for a given target.

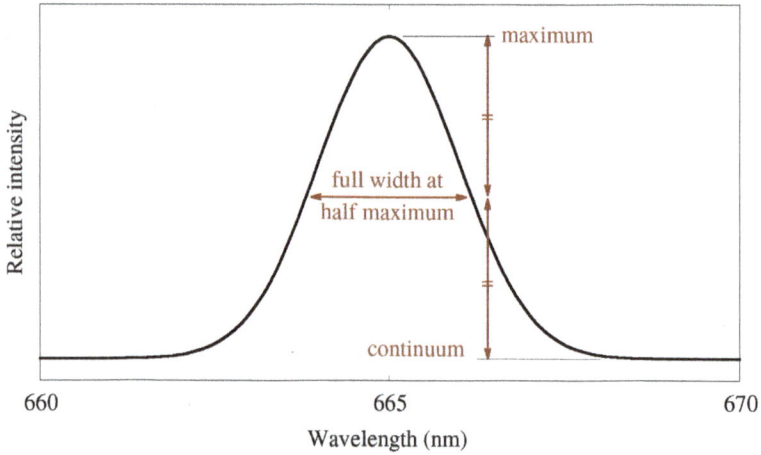

FIG. 6.3 – Definition of the full width half maximum (FWHM).

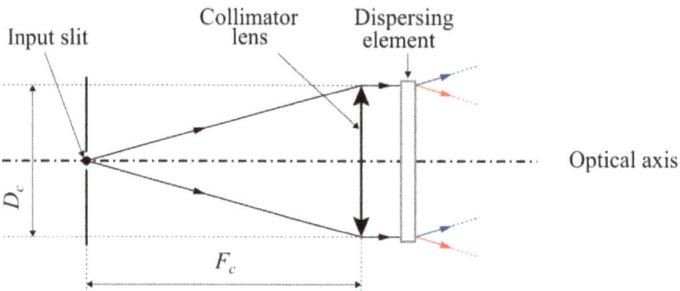

FIG. 6.4 – Definition of the F/D ratio for a spectroscope.

Beyond this first level, the focal ratio has noticeable secondary effects for telescope adaptation.

First of all, let me be precise about the definition of the spectroscope focal ratio. It is the ratio of the focal length of the collimator F_c and its diameter D_c F_c/D_c (fig. 6.4).

It is exactly the same as the F/D ratio of a telescope, but with the difference that in the telescope it is about the outgoing beam, while in a spectroscope it is about the incoming beam.

Of course, in an homogeneous configuration, the F-ratios of the telescope and the spectroscope must be the same ($F_t/D_t = F_c/D_c$). In this case, the

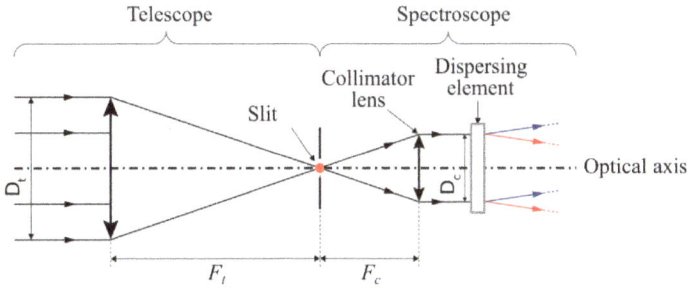

FIG. 6.5 – Optimal arrangement between the telescope and the spectroscope.

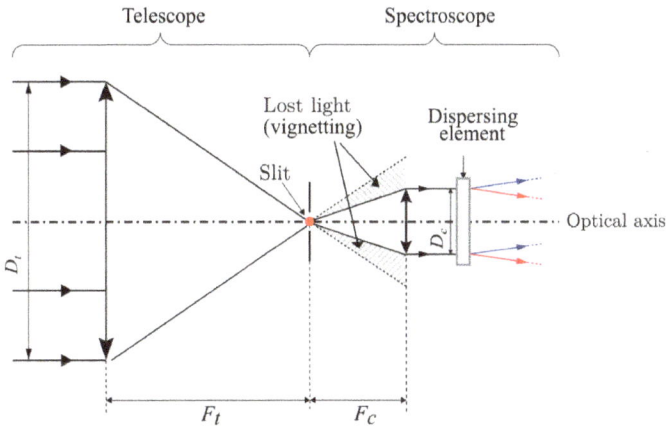

FIG. 6.6 – Vignetting in the spectroscope entry.

entire outgoing beam from the telescope can be used by the spectroscope (fig. 6.5).

If the telescope focal ratio is larger than the spectroscope one – i.e. if $F_t/D_t < F_c/D_c$ – a significant fraction of the light collected by the telescope is lost: it will not get into the spectroscope (so-called vignetting) (fig. 6.6), and you will under-utilize your telescope.

However, the spectroscope will be functioning normally, and the results you will get are definitely usable. But, it will be like putting a diaphragm on the telescope: you would have got the exact same results with a smaller telescope.

Conversely, if the telescope has a larger F-ratio than the spectroscope – i.e. if $F_t/D_t > F_c/D_c$ – then there is no light leakage (fig. 6.7).

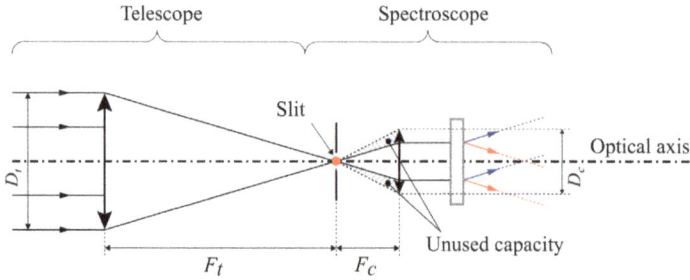

FIG. 6.7 – Telescope of too small focal ratio.

Such configuration is useable in reality, but it is far from optimal: you do not use the full capacity of the spectroscope. By putting a focal reducer behind the telescope, you effectively increase the beam focal ratio, and you are able to better exploit the focal ratio of the spectroscope. I can hear the question already: "why use a reducer, if the spectroscope can work without it ?".

Recall that there is a slit at the spectroscope entry, and it determines the resolution of the instrument. This slit is typically of the same order (a few micrometers) of the size of the image of a star at the telescope exit. Now, if you put a reducer behind the telescope, not only do you better exploit the full aperture of the spectroscope but you also reduce the size of this image, and you will get more light more easily into the spectroscope. In practice, this constraint on the slit is often critical. So much so that working with a spectroscope and not using its full focal ratio is a waste. This would mean you are losing light (it's a rare and precious thing!), or that you could improve the resolution of your instrument without any compromises.

It is thus clear that, for any given resolution, the larger the spectroscope focal ratio, the better. But this has a direct cost: to get a better focal ratio, you need larger optics and stronger optical corrections.

6.3 Magnification and Sampling

Until now, I have barely dealt with the image detector, the CCD camera. However, at this stage, it is the most essential element, since it is on this detector that the image is formed.

A CCD detector is a pixel matrix of given size (fig. 6.8). Each pixel is an independent light detector.

The pixel is thus the size of the smallest detail visible in the image. We also saw that the spectroscope slit determines the size of the smallest detail visible in the spectrum: it is clear that these two dimensions, of the slit and of

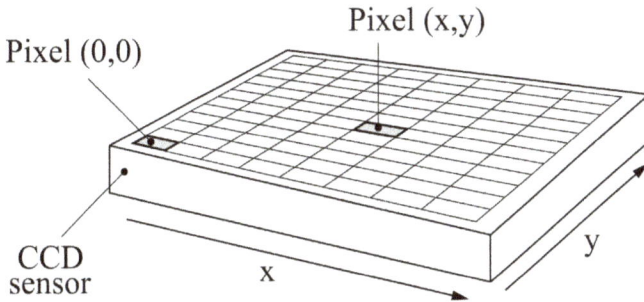

FIG. 6.8 – The CCD is a matrix of pixels.

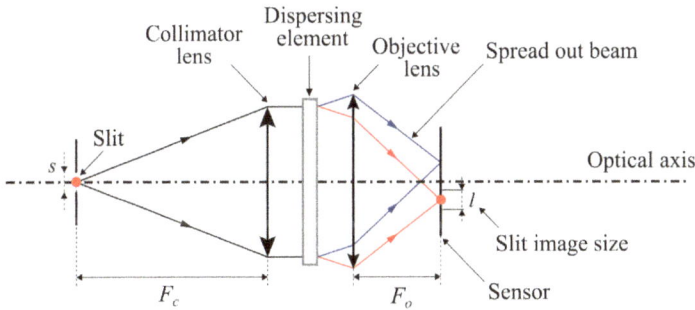

FIG. 6.9 – Spectroscope magnification.

the pixels, should be proportional. We call this proportion *sampling*, i.e. the number of pixels covering the width of the slit image size.

There is an important theorem for signal processing (the Nyquist-Shannon theorem), which says that for an optimal detection, the smallest detail visible in the spectrum should span at least two pixels. In practice, we aim – if possible – at 2.5 or 3 pixels.

We generally know the size of the slit s, but not necessarily the size of its image on the CCD, l. The latter is calculated accounting for the spectroscope magnification, given by the ratio F_o/F_c of the focal length of the objective F_o to the focal length of the collimator F_c (fig. 6.9).

$$l = s \times \frac{F_o}{F_c}$$

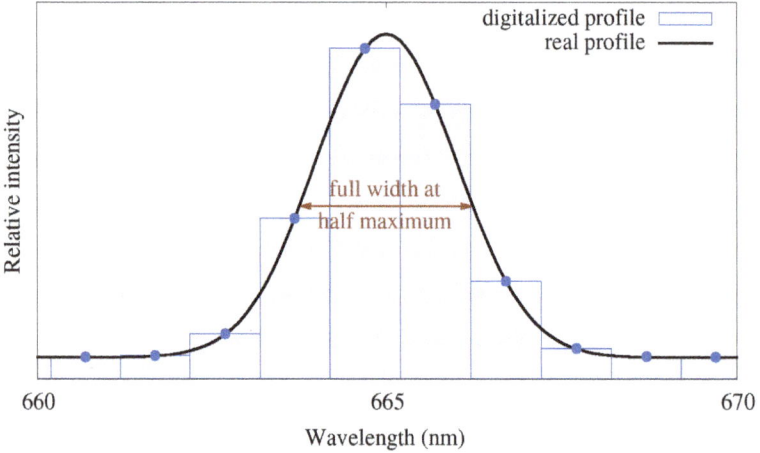

FIG. 6.10 – The right sampling.

Let's take an example. In a spectroscope[20], the slit size s is 35 micrometers, the focal length of the collimator is $F_c = 130$ mm, and the focal length of the objective is 85 mm. Moreover, the pixels on the camera are of 9 micrometers. In this case, the image of the slit on the CCD has width l:

$$l = s \times \frac{F_o}{F_c} = 35 \times \frac{85}{130} \approx 23 \, \text{micrometers}$$

Given the pixel size of 9 micrometers, the slit image spans $23/9 = 2.55$ pixels. It is perfect sampling: neither too small, nor too big (according to the Nyquist-Shannon theorem) (fig. 6.10).

You might come across less optimistic cases. With a different configuration, you can be faced with either an over-sampling, or a sub-optimal sampling.

In the case of over-sampling, the smallest visible detail in the spectrum spans many pixels (fig. 6.11).

It is not the worse case scenario, but there still are losses on multiple levels:

– Light is spread over too many pixels: the exposure time to get a sufficient signal increases;

– We have to read out many pixels to get the whole information;

– The resulting image is larger without containing more information;

---

[20] The parameters listed here are for a LISA spectroscope.

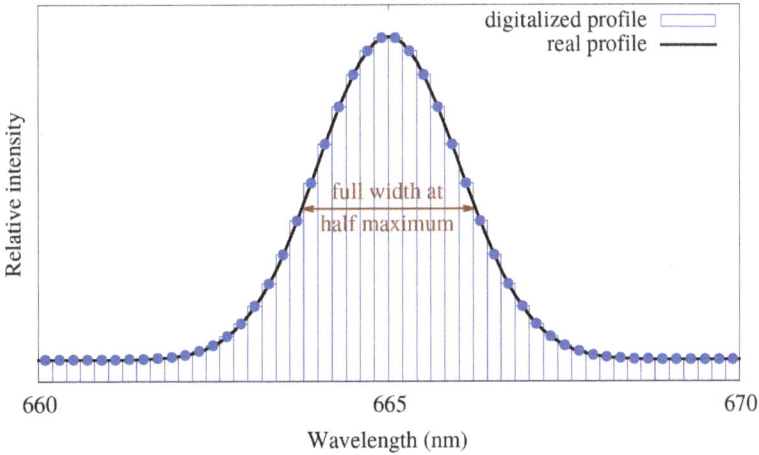

FIG. 6.11 – Over-sampling.

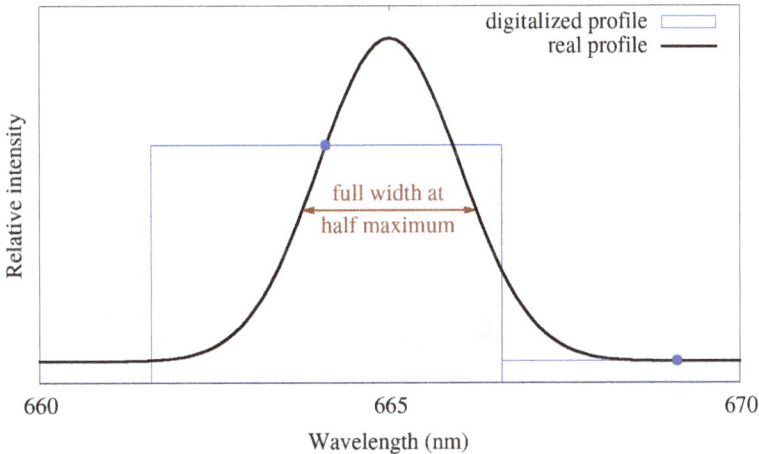

FIG. 6.12 – Sub-optimal sampling.

The sub-optimal sampling is worse: the camera limits the visible details in the spectrum. The digitised information has significantly poorer resolution than the actual spectral profile (fig. 6.12). In an extreme case, we can imagine that two lines separated by the spectroscope are blended in one single pixel by the detector.

In this case, the spectrum produced by the spectroscope is modified by the camera. This situation should always be avoided, because it is impossible to distinguish what comes from the light source and what comes from the instrument.

6.4 Resolution and Dispersion

Let me spend some time on these two often confused terms, despite being two very different quantities.

I have previously defined the resolution: it is the smallest detail visible in a spectrum, expressed in units of wavelength (nm). If the instrument is well designed and tuned, the smallest element fits on a few pixels.

The dispersion is expressed in wavelengths per unit of physical length on the detector. This can mean nm mm^{-1} (the so-called plate-factor) or nm pixel^{-1} (for a given camera).

These two quantities (resolution and dispersion) are evidently related – because of the sampling. More precisely, the sampling is the ratio between the resolution and the dispersion:

$$Sampling = \frac{Resolution \text{ (nm)}}{Dispersion \text{ (nm/pixel)}} \text{ (in pixels)}$$

When the sampling is right (about 2.5 to 3 pixels), talking about resolution or dispersion is equivalent. But there is a common mistake, which consists in evaluating the dispersion of a spectroscope starting from the grating equation, and then infer the resolution using the size of the pixels. This is wrong, because the resolution depends on the optics of the instrument, not on the pixels size.

6.5 Spectral Range

We saw that, depending on the spectroscope resolution, the image of the spectrum is more or less stretched on the CCD. But the CCD does not have an infinite length, and for a given detector size, the larger the resolution, the smaller the spectral range. We call the spectral range the bandwidth covered by the detector. Figure 6.13 shows a spectral bandwidth of 50 nm around the line H_α (between 625 nm and 675 nm).

Of course, the larger the detector, the larger the spectral range – but even here there are some limits. We saw that a spectroscope has to compensate for optical aberrations to produce a well-corrected spectrum over its whole range, and the larger the range is, the harder it is to make the corrections. Thus, to increase the spectral range, we can use a detector as large as possible, but within the reasonable limits specified by the manufacturer. Beyond them, there is necessarily a quality loss in the spectrum.

For example, the spectroscope Lhires III was initially designed for a CCD of length about 8 mm. Nowadays, it is easy to find larger detectors, and it is possible to extend the spectral range of the instrument. But when the detector

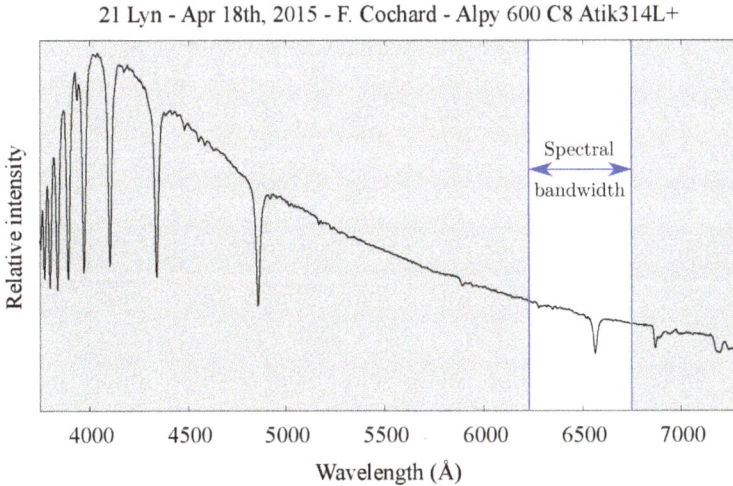

21 Lyn - Apr 18th, 2015 - F. Cochard - Alpy 600 C8 Atik314L+

FIG. 6.13 – Definition of spectral range.

is really big (say more than 12 mm), you can clearly see that the quality of the spectrum is not the same on the edges. This is particularly clear with calibration spectra.

6.6 Spectral Domain

We are often facing this dilemma: the higher the resolution, the smaller the spectral range. We need to find the best compromise for our target – some observations focus on the blue, some others around H_α, and yet some others in the near IR.

Note that this is one of the advantages of low resolution: the spectrum covers almost the entire visible range, so no choice has to be made!

Whenever the spectral range is limited, it is still possible to chose which part of the visible (or near IR) spectral domain we want to observe. In most instruments, we can make this choice by simply changing the inclination of the grating. Therefore, we have another parameter of the spectroscope, complimentary to the spectral range: in which wavelength interval can we use this range? Figure 6.14 shows for example two spectral domains. The first is around the line H_β and the second is around the line H_α.

6.7 Efficiency

Another parameter of the spectroscope is its efficiency, i.e. the number of photons of given wavelength which will reach the CCD divided by the number of photon entering the slit. This efficiency as a function of wavelength depends

21 Lyn - Apr 18th, 2015 - F. Cochard - Alpy 600 C8 Atik314L+

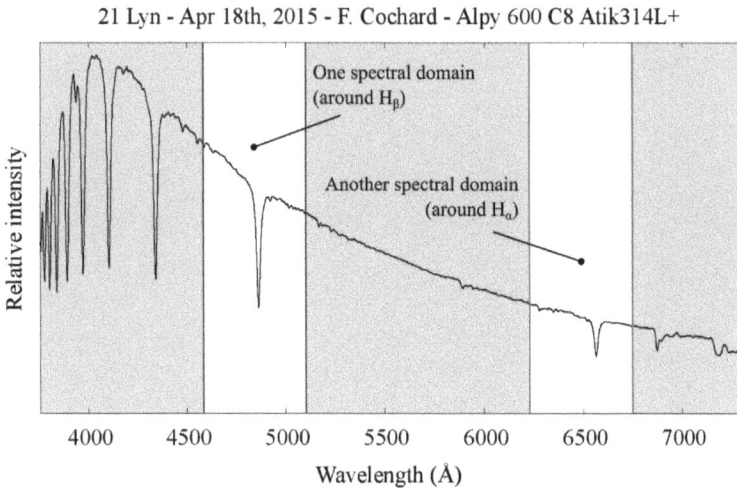

FIG. 6.14 – Two different spectral domains.

on many factors: number and quality of the optics, internal vignetting, grating efficiency... Moreover, it is a complicated parameter to estimate, since you need to do exactly the same measurement with and without the spectroscope properly isolating each wavelength.

The efficiency of the spectroscope is also impacted by two main external factors:

– the fraction of the stellar light entering in the slit;

– the efficiency of the CCD (and the camera electronics in general).

It is easily seen that the ensemble of these elements is paramount for our observations: if I double the efficiency of my instrument, I can reduce by a factor of two the exposure time, without compromising the quality of the results.

Overall, when considering the ensemble of the instrument (including telescope and CCD camera), the numbers are worrying: even in professional instruments, an overall efficiency of less than 10% is common!

Between two otherwise identical instruments, you should always favor the one with higher efficiency...but in practice, it is difficult to have solid grounds for this issue and make an informed choice. In practice, I've never seen a choice being made for this reason only. Of course, great attention must be focused on this during the design of an instrument (choice of the dispersion element, of the instrument architecture, and of the camera). But once these choices are made, the observer has only one responsibility: you need to inject as much light as possible into the slit – and my experience says that you can lose 90%

of the light there. I will come back extensively to this in the part dedicated to observation.

6.8   Mechanical Backfocus and Fastening

The spectroscope is put between the telescope and the acquisition camera. You need to make sure that the mechanical anchoring and optical alignment between the three elements (telescope, spectroscope, and camera) is optimal. Pay particular attention for the rigidity of the mechanical support: as the telescope moves along the sky, the smallest mechanical flexure results in a relative motion of the spectrum on the detector of a few tenths of millimeters, with catastrophic consequences (loss of resolution, loss of calibration, etc.).

Input Backfocus: Adaptation to the Telescope

The telescope makes an image on its focal plane. This plane is outside the telescope, to allow one to put a detector on it (eyepiece, CCD camera, ...). We call *backfocus* the distance between the end of the telescope and the focal plane (fig. 6.15). This distance is variable thanks to the telescope focusing system. We thus have a range of possible backfocuses.

The slit of the spectroscope needs to be positioned on the focal plane of the telescope. It is inside the spectroscope, at a distance from the outer edge of the spectroscope also called backfocus – the input backfocus of the spectroscope (fig. 6.16).

The input backfocus of the spectroscope needs to be compatible with the range of the telescope backfocuses. If the telescope backfocus is too long, the problem is easily solved: we can add extension rings. On the other hand, if the output backfocus of the telescope is too small – as it can be for telescopes with small focal ratios, such as Newtonian telescopes – then the problem is more complex. One can use a Barlow lens to increase the focal length of the telescope: this pushes the position of the focus rearward. But such device also

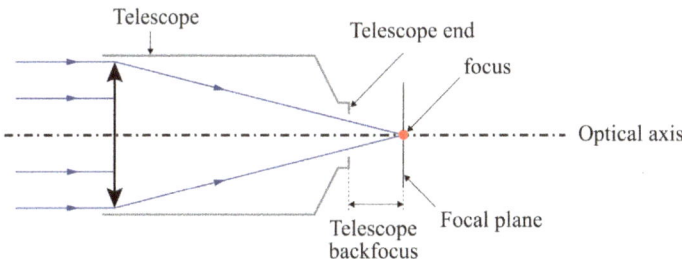

FIG. 6.15 – Output backfocus of the telescope.

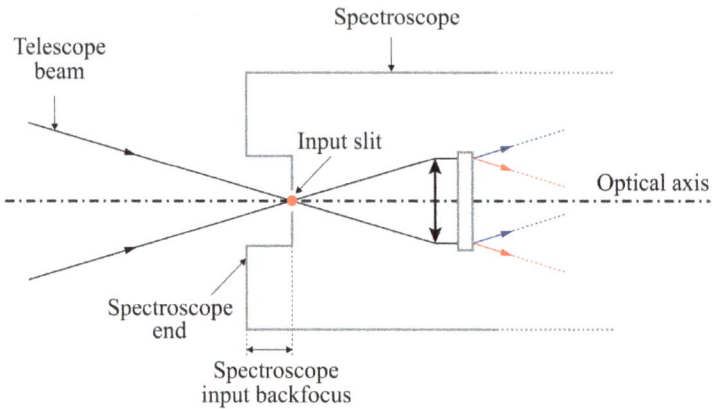

FIG. 6.16 – Input backfocus of the spectroscope.

modifies the focal ratio (F/D) of the telescope: make sure it still matches your spectroscope.

Also look at the support for the spectroscope on the telescope: it is either a sliding system (of 2" (50.8 mm) diameter), or a system with threaded rings (Schmidt-Cassegrain telescopes often have a threaded ring of 2" diameter × 24 tpi). Think also, that when you put your spectroscope on the telescope, you need to orient it in the right way (the same way we orient a camera for imaging). The support must allow for the rotation of the spectroscope.

Output Backfocus: Adaptation to the Camera

At the other end, the spectroscope forms an image of the spectrum, at a distance called *output backfocus* (fig. 6.17). The spectroscope allows adjustment of this backfocus in a limited range. Depending on the instrument, this range can be very short (for example, with the Lhires III the CCD must be within 2 mm – because of the Littrow architecture of this instrument); for others this range is more generous (of the order of 10 mm for the Alpy 600).

The camera also has an input backfocus (fig. 6.18), which needs to be compatible with the spectroscope. This is not always easy: some cameras have a large backfocus range (greater than 35 mm), for example when they have a filter wheel. A DSLR with a T ring (thread $M42$ with pitch of 0.75 mm) can reach almost 55 mm of backfocus. The situation is similar to what we saw before (adaptation to the telescope): we can add extension rings (T-extender), and in some cases we can put a Barlow lens on the spectroscope to increase it's output backfocus (as in the case of the Alpy 600, which becomes compatible with DSLRs in this way).

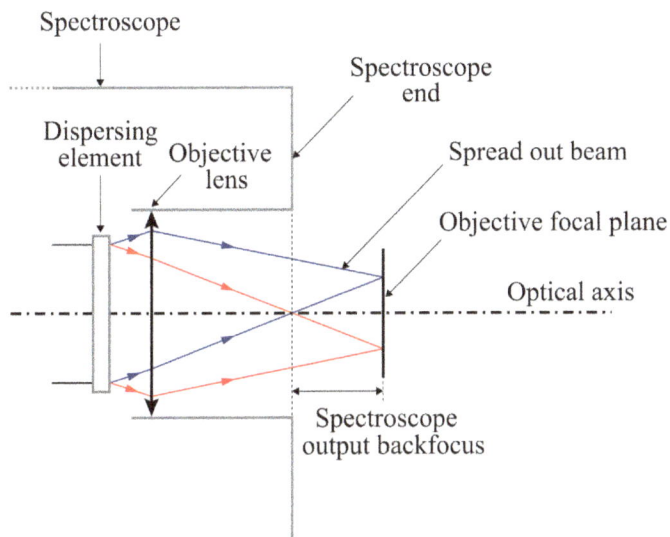

FIG. 6.17 – Output backfocus of the spectroscope.

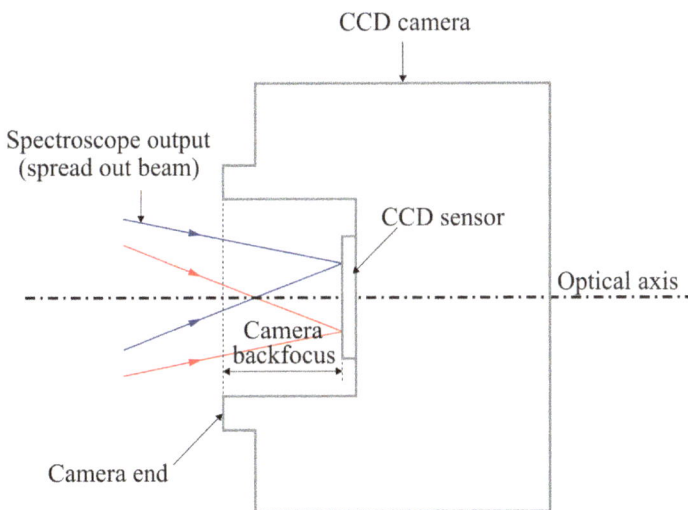

FIG. 6.18 – Camera backfocus.

Most cameras have nowadays a T-mount thread.

Make sure the camera adapter allows you to orient it, so that the spectrum is perfectly horizontal in your image.

6.9 Telescope and Sky Quality

After looking at all the parameters, you need to verify that your different elements are "coherent". But before checking this, we still need an important element: the size of the image of a star produced by the telescope.

A star, despite its huge size, is seen from Earth as a point. Thus, it should give an infinitesimally small image in the focus of the telescope. This is not the case for at least three reasons:

- the light from the star goes through the Earth's atmosphere, which has a certain turbulence. This turbulence deforms the point-like image of the star, and makes it scintillate. The immediate result is that the image of a star in the focus of the telescope is not a point, but a small blur;

- any telescope is diffraction limited – it is not a negotiable optical law. If your telescope is in an area where the atmospheric turbulence is negligible, the image of the star is formed by a central disc surrounded by a few rings. The bigger the telescope, the smaller the diffraction pattern;

- Perfect optics do not exist, and even the best instruments have optical aberrations.

Usually, the atmospheric turbulence is dominant, and the diffraction limit is very rarely reached. The turbulence changes as a function of the weather, but it is considered an intrinsic parameter of the observing site. When professional astronomers look for the best observing sites in the world, this is the first criterion they consider.

The turbulence of an observing site is characterized by the *seeing*. It is a small angle, corresponding to the angular diameter observed for a star. The seeing is measured in arcseconds. A very ordinary (or even mediocre) site can have a seeing of a few arcseconds; a very good site can go below an arcsecond.

Since we can think of a telescope as a simple lens, it "converts" this angular diameter (expressed in arcseconds) into a real diameter (in micrometers) at the focus plane. Figure 6.19 shows this process in a simplified way.

The seeing angle α is typically very small, therefore, we can equate the angle with its tangent, if we express the angle in radians. Consequently, the seeing α (expressed in ", or arcseconds), the size of the image at the focus i (in micrometers) and the focal length of the telescope F_T (in meters) are related by:

$$i = F_T \times \alpha \times \frac{\pi \times 10^6}{3,600 \times 180} \approx F_T \times \alpha \times 4.85$$

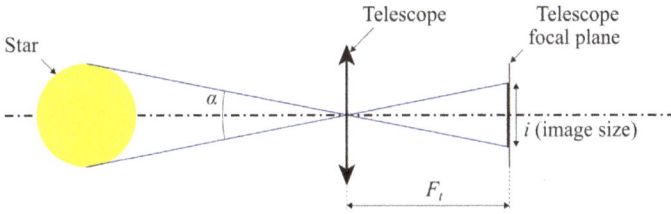

FIG. 6.19 – Angular diameter and image size.

Let's take an example. If your sky has a seeing of 3 arcseconds and your telescope has a focal length of 2 meters, then the image blur at the focus is of size $i = 2 \times 3 \times 4.853 \approx 30$ micrometers.

This is an order of magnitude to keep in mind for the rest: the size of the image of a star from a small telescope pointed to an ordinary sky is of the order of 20 to 30 micrometers. This is also the most common size for the slit of the spectroscope.

I want to emphasize that the size of the image does not depend on the magnitude of the star. We are used to seeing the brightest stars bigger than the fainter stars in an image of the sky. This is just the effect of the visualization threshold of the image. If you measure the full width half maximum (FWHM) of each star, you will find very similar values (I encourage you to do the exercise on your own).

The size of a star at the focus is proportional to the focal length of the telescope. Double the focal length, and you will have stars twice as big. This must be kept in mind when you put a slit at the focus, as the entry point of the spectroscope. The focal length of your telescope can be modified; in one direction with a focal reducer, and in the other with a Barlow lens (often multiplying the focal length by a factor of 2 or 3). Of course, these accessories also modify the size of the star at the focal point.

To measure the seeing, it is "sufficient" to take an image of stars for a few seconds – on a perfectly set up telescope – and to measure the size of the stars on the image (you measure the full width half maximum). The duration of this measurement allows you to average the turbulence; if the exposure is too short, the turbulence is not frozen.

The resulting size i (expressed either in pixels or micrometers) is the size of the star at the focus; the seeing can be obtained with the following relation:

$$\alpha \approx \frac{i}{F_T \times 4.85}$$

(in arcseconds, if i is in micrometers and the telescope focal length F_T is in meters).

6.10   Adjusting the Configuration

Now we have all we need to verify the capability of our instrument. Let's start again from the conceptual scheme of the spectroscope (fig. 6.20).

We can consider our setup coherent if the three following conditions are satisfied:

– the size of a star at the focus is comparable to the slit size;

– the size of the image of the slit (including the magnification) gives a sampling slightly larger than 2 pixels;

– the F/D ratio of the telescope is comparable to the focal ratio of the spectroscope.

Moreover, you have to check that the resolution and the spectral range given by this configuration are consistant with the observation you want to perform.

Is all of this too complicated? No, because in the end the things we can adjust are limited: typically we can change the slit size (often spectroscopes offer several slits), and the adaptation of the focal length and the focal ratio with a focal reducer or a Barlow lens.

The rest depends essentially only on the telescope diameter. For a given focal ratio, if you increase the telescope diameter, you also increase the size of the image of a star in the focus. Thus, to collect all the light into the spectroscope, you also need to increase the slit size – and if you want to maintain the same resolution, you *necessarily* need to proportionally increase the size of the spectroscope!

This is a good news for amateurs: since we work on small telescopes (compared to professional instruments), we can have the same resolution with significantly smaller spectroscopes.

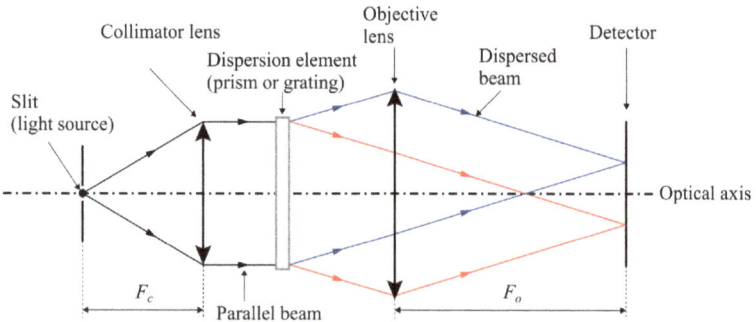

FIG. 6.20 – Spectroscope architecture.

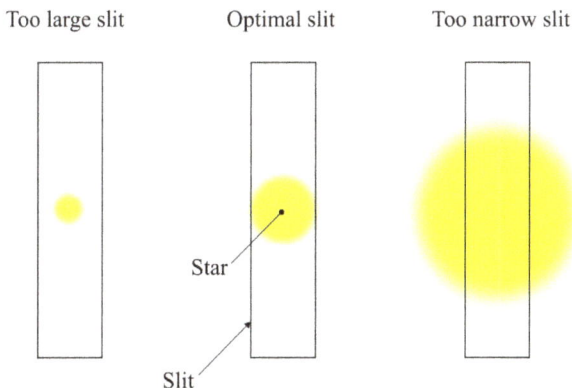

FIG. 6.21 – Adaptation of the slit size.

So, to summarize, if you work with a small telescope – say up to 300 mm of diameter – you will not have difficulty adapting it to a commercial spectroscope. If you think you will be working with larger telescopes, you need to consider carefully the adaptation issue and make some necessary compromises.

Before closing this chapter on the instrumental parameters, I still have a few words about the two adaptation possibles: the size of the slit and the focal length of the telescope.

How to Chose the Slit Size

In most of the instruments manufactured by Shelyak Instruments, the observer can chose between different slit sizes. This adds flexibility to observing, but it requires a competent choice.

We have seen that the slit size needs to be proportioned to the size of a star image at the focus plane, i.e. to the telescope focal length and the site seeing. If the slit is too small, part of the light from the star is lost (it does not fit into the slit). Conversely, if the slit is too big, the star has room to move around within the slit: this is a pity, because using a smaller slit increases the resolution without any sacrifice (fig. 6.21).

To make the right choice, we need to consider the following two constraints:

– size of the image of the star in the focal plane of the telescope (slit plane);

– sampling of the spectrum.

If I use a bigger slit, I collect more light (thus reducing the exposure time, etc.). However, if I reduce the slit size, I can increase the resolution – as long as I maintain a sufficient sampling (larger than 2.5-3 pixels).

In practice, there is no issue for small instruments (say, diameters smaller than 300 mm): in these cases the size of the star and the size of the pixels are naturally in the right proportion. Things slightly change for larger instruments (diameters above 300 mm): in these cases the slit corresponding to the correct sampling (typically of 20-25 micrometers) is small compared to the dimension of a star on the focal plane.

Recall that the dimension of the star depends only on the telescope focal length and the seeing.

Then, it might be convenient to use a larger slit to better exploit the telescope and reach fainter magnitudes (or reduce the exposure times). But this necessarily implies a reduced resolution and some over-sampling.

To summarize, if you have a large telescope, you can think of varying the slit size, but you will need to find a compromise between the resolution and the luminosity of the targets.

With all I have said until now, you might be thinking that you continuously need to adjust the slit to the observing conditions (there are systems with interchangeable slits). Luckily this is not the case: the elements listed here are useful to choose your configuration, but once chosen, you don't modify it anymore. This is also relevant for the other parameters: since the slit is fixed, the resolution of your spectra is an intrinsic parameter of your instrument and does not depend on the observational conditions anymore. If you observe during a night with a very turbulent sky, you only see a portion of the starlight crossing the slit (and thus the exposure time to reach the same amount of light in the instrument will go up), but the resolution will stay constant.

Adaptation of the Telescope Focal Length

We can adapt the focal length of the telescope and its focal ratio by adding a focal reducer or a Barlow lens. Let's have a closer look on how this works.

These accessories are purely optical elements that modify the beam exiting the telescope and the position of the focus (fig. 6.22).

Consequently, the effective focal length of the instrument is either shortened (focal reducer), or increased (Barlow lens). Each accessory has its own coefficient by which it multiplies the focal length (smaller than 1 for a reducer and larger than 1 for a Barlow lens).

We can modify the focal length, but in any case, the diameter of the telescope remains unchanged. Consequently, the F/D ratio of the telescope is modified exactly like the focal length (same multiplying factor).

Let me take an example. I have a telescope with focal length 200 mm and with focal ratio $F/10$. The nominal focal length is thus $F = 200 \times 10 = 2,000$ mm.

- If I put a "common" focal reducer of $\times 0.63$, the effective focal length becomes $F_{eff} = 1,260$ mm, and thus the focal ratio becomes $F_{eff}/D = 1,260/200 = 6.3$ (i.e. 10×0.63).

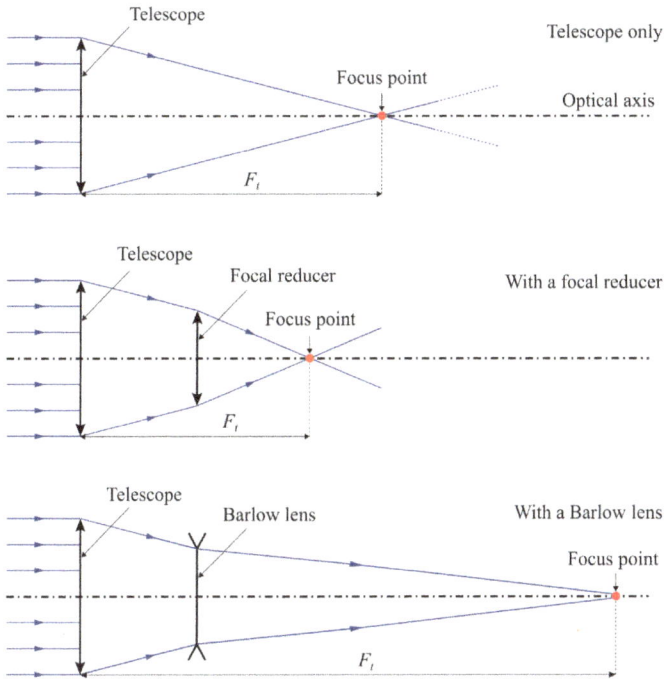

FIG. 6.22 – Focal reducer and Barlow lens.

– If I put a focal doubler (Barlow lens ×2), the effective focal length becomes $F_{eff} = 4{,}000$ mm, and thus the focal ratio becomes $F_{eff}/D = 4{,}000/200 = 20$ (i.e. 10×2).

We cannot distinguish these two effects – modification of the focal length and modification of the focal ratio. I remind you that the focal length determines the size of the star at the focus, and the focal ratio needs to fit the focal ratio of the spectroscope: the room for adjustment is small.

I want to underline a very important secondary effect of this adjustment. We saw that the optical aberrations are increasing with the beam focal ratio. Increasing the focal length of an instrument is usually not a problem, since the resulting beam is narrower than the original. On the other hand, the focal reducer spreads the beam, increasing the optical aberrations. If the reducer is designed explicitly for your telescope, things should go well. But if you match any random reducer on your telescope, it is likely to significantly downgrade the image quality on the focal plane. In this latter case, you probably lose the advantages of this operation.

Be aware that the focal reducers are designed to work in very specific conditions (distance from the focal plane), for the same exact concerns about the optical aberrations. If you don't respect these conditions (usually indicated by the maker), you will create optical defects that will later affect your spectrum.

If you use a modest instrument ($D \leq 300\,\text{mm}$), you can bet on a very smooth and natural adaptation, since these instruments produce image sizes of the stars that are close to the size of pixels of current cameras.

Only if you work on instruments with larger diameters you might need to choose between the resolution (using a smaller slit, ideal for the spectroscope) or the luminosity (using a larger slit, ideal for the telescope). You can choose the high resolution for some observations (details of the line profiles, velocity measurements with Doppler shifts), and then switch to observe a fainter object at lower resolution.

6.11 Keep it Simple!

I have tried to give you all the keys to define the best configuration depending on your equipment. This might seem complicated if you are not familiar with these optical elements. But, a little experience will show you that it's easy to adjust them!

In the meanwhile, do not get hung up on these elements. These are needed to optimize an instrument, but there is no need for an optimal instrument to begin in spectroscopy. I repeat that, in any case, the quality of a spectrum is intrinsic to the spectroscope, and the optimization can only improve the efficiency. To begin, you will observe bright stars, and if you need an exposure of 5 seconds instead of 0.1, it is not a big deal.

I urge you to work with *your* instrument, even if you know it is not optimal. It is the instrument you know, and you already have. It is always better to observe now with the available means than postponing your first steps in spectroscopy. The first observations will give you significant progress, and you will be in a better position to judge the eventual limits of your instrumentation.

Chapter 7

CCD Cameras and Acquisition Softwares

CCD cameras and their popularization are the heart of the recent developments in astronomy, especially in the amateurs' domain. It is thanks to the availability of high-performance CCD cameras and their acquisition software packages that spectroscopy can develop today.

There are many books talking in detail about these detectors and my objective here is not completeness. I will only focus on the specific elements of use in spectroscopy.

If you are completely new to practical astronomy, including CCD cameras, I invite you to spend some time doing deep sky imaging before going to spectroscopy. Be it the telescope handling, or the use of the camera (image acquisition, pre-processing), anything you learn will necessarily and entirely be re-used in spectroscopy.

Excluding some peculiar cases, we need at least two cameras for spectroscopy: one for the acquisition of spectra, and one for guiding. It might be useful to use a third camera for telescope pointing.

7.1 A Wide Variety of Choices

We are lucky to have a large variety of possible choices for CCD cameras with varying functions and prices. I will review the most important characteristics for the cameras we need.

The expression CCD *camera* can be used to indicate both devices to take pictures and record videos. In our case, we need to take single snapshots, so we use it in the former sense only.

Acquisition Camera

It is the most important camera of the whole system. The criteria to consider are the following:

- Size of the detector. The bigger the better (you can always fit a smaller image on a larger detector). Typically, your budget is the limiting factor. Although, beware of very large detectors: they create large image files and the read out time can become prohibitive. In these cases, you need to be able to crop the image (to read only part of it). For our application in spectroscopy, it is sufficient to choose a detector which fits the whole useful spectrum (with some buffer space);

- Pixel size: it is the parameter determining sampling. The bigger the pixels are, the more sensitive to light – for any given intrinsic performance level of the material they are made of. Thus, may be tempted to look for find the biggest pixels available – which is tricky because the tendency of the market is to make them smaller to reduce the costs. The most important thing to check is that the size of the pixels allows for the right sampling with your instrument (see section 6.3);

- Read out dynamics. It is the number of gray levels that the camera can read out of each pixel. It is usually expressed in number of bits of the camera processor. A 12 bits camera offers $2^{12} = 4,096$ shades of gray. Nowadays 16 bits cameras are common ($2^{16} = 65,536$ shades of gray);

- Quantum efficiency. It is the camera's sensitivity to light (i.e. its capacity to detect photons). The higher the better but also, typically, the more expensive the camera is;

- Read out noise. It is a parameter depending on the quality of the read out electronics of the camera, which is measured in electrons (e^-). The smaller it is, the better. Note that this read out noise, which is a "reading" error, can be reduced in some cameras by slowing down the read out process;

- Cooling and dark current. The CCD detector has a finite temperature, which means that, even in the total absence of light, there will regularly be electrons detected that did not arise from the arrival of a photon. This is a systematic effect, which we compensate during the pre-processing. Effective cooling of the camera (at least 20 degrees below ambient temperature) can strongly reduce this effect, allowing for long exposure times without losing too much of the available dynamical range;

- Shutter. The camera must allow for very short exposure times (tenths of seconds, possibly less) for those rare cases when light is abundant (Solar spectra, lamp spectra, very bright stars), but in most cases the exposure time will be several seconds, or even minutes. Present day cameras do not have a mechanical shutter (they are replaced by an electronic system): this allows for very short exposures.

Color Detector?

Many manufacturers offer color cameras. One might think they are particularly suitable for spectroscopy – whose essence is the study of colors – but that is a mistake. A color CCD detector has several disadvantages for scientific exploitation:

- each pixel has a red, green, or blue filter and the combination of different "colored" pixels allows one to determine a color for each of them (the decomposition of colors in pixels is done using a Bayer matrix). The consequence is that at best one pixel out of three sees some light in any given portion of the spectrum – which is a clear limitation for efficiency;

- to balance the "holes" left in the spectrum by the pixels with the wrong filters, one needs to over sample, which leads to an extra efficiency loss;

- the image is much harder to process, since one needs to take into account the Bayer matrix.

For these reasons, most of the detectors used in scientific applications are monochromatic, recording in gray shades – and in the context of spectroscopy this is not disturbing at all because the color variations are transformed to variation in the position of the signal on the detector.

Can I use a Digital Camera (DSLR)?

It is possible to use a DSLR as the CCD detector. It is not the most efficient way of doing spectroscopy, but these devices are very common among many observers. Therefore, they are a good way of getting started, and when we design our spectroscopes, we always make sure that they can be used with a DSLR.

However, let me be clear: if you like spectroscopy, you will have to go to a CCD camera intended for astronomical use. They have higher performance (better sensitivity, cooling) and are easier to use (no Bayer matrix, standard format, and native digital control from the computer).

This simplicity is important, even for beginners: the image from a CCD camera is immediately usable with imaging software, while the image from a DSLR camera needs to be converted from its native format (RAW) into a standard astronomical format (FITS). It can be done, but it is a tedious operation that limits your productivity.

Guiding Camera

The requirements for the guiding camera are less stringent than those for the main acquisition camera. The reason is simple: this camera does not contribute directly to the quality of the spectrum. Nevertheless, you should not use just any camera. In cases when the target is a very faint object observed at low resolution, the guiding camera can become the limiting factor (you need to see the object to take its spectrum).

Guiding maintains the image of the target star in the slit, therefore the guiding camera needs to be able to take pictures at a high rate (a few images per second) – but there are cases when the exposure can be as long as 30 seconds to catch faint objects (magnitude 15, for example). Some guiding cameras come from the video realm rather than imaging. In this case, you need to make sure your camera is also able to have long exposure times (a few tens of seconds).

Guiding cameras often have a small detector. It is not a critical element if your installation has good pointing capability (i.e. any faint object can be positioned in the middle of the detector of the guiding camera), because, in the end, all you need is to track the position of the star in the slit. On the other hand, if your installation has mediocre pointing precision, you will need a larger detector to "catch" the star in the guiding field (I will extensively come back to this, see section 11.4). Personally, I prefer to have a third camera dedicated to pointing (electronic finder), and live with a small detector for guiding.

An important feature of the guiding camera, is that it needs to be controlled by software capable of also controlling your mount to allow for autoguiding (see section 11.5). It is probable that the software shipped with the camera is not able to do this. In reality, it is better to take the opposite approach and use a camera that your favorite software for autoguiding can control.

Pointing Camera

The pointing camera (also known as *electronic finder*), if you want to use one, has even less constraints than the previous ones. Often, an old camera will work well for this purpose. I personally use a video camera directly connected to a small screen (without using the computer at all). But you might as well just use a basic CCD camera; which also allows for automatic field recognition.

7.2   Image Detector or Light Detector?

A CCD camera is often considered as an image detector – exactly as a digital camera. The outcome of the acquisition is a file containing the image. This is all very true when doing imaging. But there is another representation possible, which I prefer when doing science: a CCD is first of all a matrix of light detectors. Each pixel can be considered an independent light detector.

The function of each physical pixel is to convert light into electrons – which is an analog process – and the role of the camera electronics is to convert the number of electrons from each pixel into a number. In the end, for every pixel, we have an associated number that represents the amount of light received by the pixel.

This number is an integer, in an arbitrary unit called ADU (which stands for *Analog to Digital Unit*). One ADU corresponds to the smallest light variation that the camera can detect; while the read out dynamic range of the camera corresponds to the maximum number of ADUs it can handle. For example, a camera with a dynamics of 16 bits will have a number of ADUs always between zero and $2^{16} = 65,536$.

Unfortunately, the capacity of each pixel to accumulate electrons is limited. This means that if the available amount of light is greater than 65,536 ADU (for a 16 bit camera), then the pixel is saturated – which means it is effectively blinded. Even if the light increases, that pixel will not provide a number of ADUs greater than this.

A CCD detector is extremely sensitive: if you expose it directly to light without a lens, it saturates in a few milliseconds!

You can shine light on a pixel well beyond its saturation point without damaging it – as soon as the amount of light drops below the maximum threshold, it starts working normally again. However, some CCDs have a latency effect, i.e. they need a significantly long time – several minutes – before recovering their normal behavior. If possible, you should avoid these detectors to make physical measurements.

The only way to avoid saturation (given a specific light source) is to reduce the exposure time.

In astronomy, it is common to experience problems with the read out dynamic range: some parts of the image are saturated or close to saturation (stars), while others are very mildly illuminated (sky background, nebulae, ...). To capture more details in the dark regions one would like to increase the exposure time – but the immediate drawback is a loss of important information in the brightest parts.

On the other hand, the exposure time should not be too short either: it is a pity if the detector can reach several thousand ADUs, and we only get maximum values of a few tens or hundreds. If this happens, you are not exploiting the full read out dynamic range of the camera (i.e. we will get very few gray levels compared to what the camera could deliver).

Let me stress this point. When you process images and spectra, you always have to have in your mind these questions:

– Is my image saturated or close to saturation?

– Is the read out dynamic range of the camera fully utilized?

– Is the exposure time suitable to what I am looking for?

Visualization Thresholds

When you look at an image, you immediately spot if there are white regions, potentially saturated. Beware, there is a trap you will necessarily face when processing your images! In reality, the read out dynamic range of the CCD (typically around 16 bits) is much larger than what a computer screen can show, and than the human eye can perceive. Let me clarify: the human eye is sensitive to a huge light range (or order 2^{25} levels), but only through adaptation of the iris (pupil dilation).

If we want to see details in an image – and in spectroscopy all the information is in the details – we need to adapt the visualization threshold of to the screen.

To adapt the threshold means deciding which (ADU) value of the image corresponds to black on the screen, and which value corresponds to white. An example: consider this image of the Andromeda galaxy ($M31$) with a dynamic range of 50,000 ADU. If I show the image with a low threshold (between 1,000 and 5,000 ADU – fig. 7.1), the arms of the galaxy are well visible, and the entire background seems grayish. On the other hand, if I chose a high threshold, say between 10,000 and 15,000 ADU (fig. 7.2), then only the core of the galaxy and the brightest stars are visible. In these two cases the visual result is very different, while the image is exactly the same.

On your DSLR, you probably have a "histogram" function, which allows one to see at a glance the intensity distribution of the image. It is a handy way to verify that the exposure is right: you can see saturated pixels and whether you have used the read out dynamic range available.

The take home point is that it's not because part of the image is white that it is saturated. Conversely, it might be saturated even if it is not white! In practice, you need software which can read the value of each pixel, the maximum and/or average level of a region of the image, etc. These are basic functions for any astrophysical software.

An Almost Perfect Detector

Pixels are a fantastic thing – especially when you can pack several millions of them in a few square millimeters. We can simply consider that the value of each pixel is the measurement of the amount of light detected; which is what we do when we look at a digital image. But at closer inspection, such measurement is affected by several defects. These are inherently caused by the manufacturing process of the detector and the camera, and are unavoidable.

The situation is not desperate though, and we can compensate for most of these defects in a certain number of operations (the so-called pre-processing).

The main defects are the following:

– the bias. If the value produced by the detector is proportional to the amount of light received, one would expect a null value when the detector is in absolute darkness. This is not true because of the electronics.

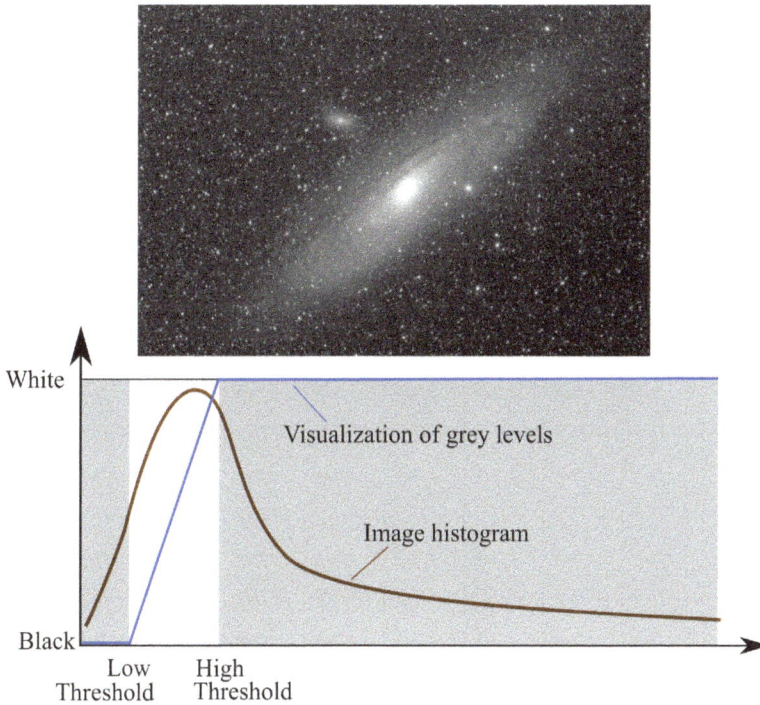

FIG. 7.1 – Visualization of the image of $M31$ with low visualization threshold.

Consequently, the value associated with a pixel in complete darkness is between a few ten to a few thousands of ADUs;

– the dark current. We have seen that the intrinsic thermal activity of the detector is interpreted as light. The longer the exposure time, the higher the thermal noise is, even in total darkness;

– the read-out noise. The whole analog-electronic process translating the number of electrons collected into ADUs has its own defects: the transfers internal to the CCD, the electronic transfers, the amplification, and finally the analog-digital converter. At the end of the procedure, the result is characterized by a quantified uncertainty;

– each pixel is not exactly as sensitive as all its neighbors. The differences are small, and relatively easy to compensate for;

– the spectral response. The CCD detector is not as sensitive at all wavelengths. The manufacturers give a spectral response curve, which is usually bell-shaped and centered between 370 and 1000 mm (fig. 7.3);

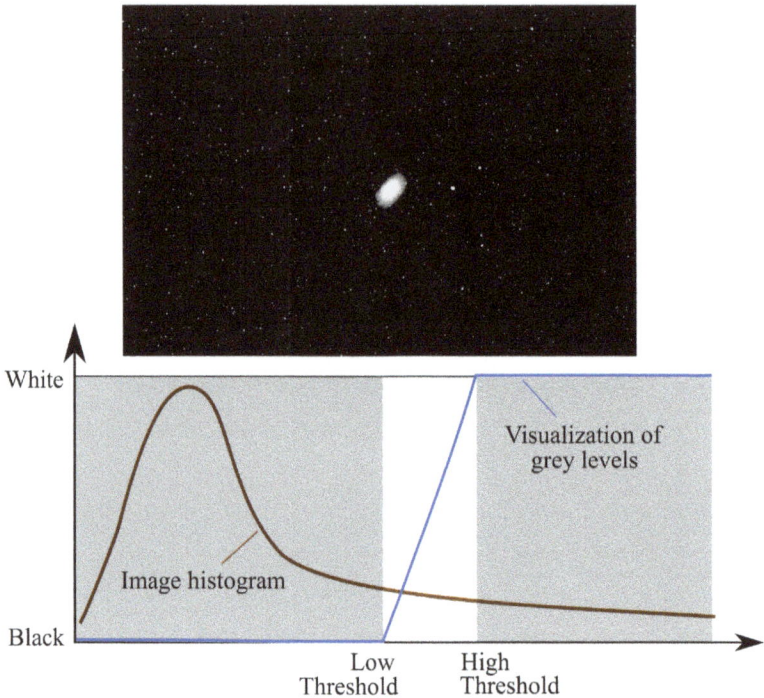

FIG. 7.2 – Visualization of the same image of *M*31 with high visualization threshold.

– linearity defects. Sometimes, with certain cameras, non-linear phenomena can be seen at high light intensity (close to saturation): the measured ADU level is not exactly linearly proportional to the amount of light. This is often found in cameras with "antiblooming" capabilities, which limit the saturation.

All these imperfections are inherent to the camera, and do not depend on the image formed on the detector. The spectroscope adds other imperfections (for example, a theoretically infinitely thin spectral line is not necessarily projected on one single pixel of the detector), and that will be covered elsewhere.

7.3 Acquisition Software

A DSLR can work independently (that is its biggest advantage), which is not the case for a CCD detector. The CCD camera is controlled with acquisition software running on a computer. The choice of acquisition software you will use for your spectroscopy observations is very important: it must not only

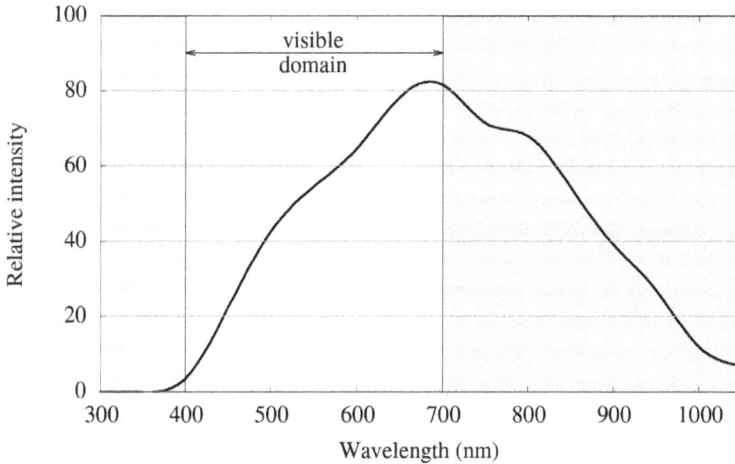

FIG. 7.3 – Typical spectral response curve for a CCD.

properly control your camera, but it also needs to provide the functionality needed for our purposes (we just reviewed some of them).

Most cameras available on the market nowadays ship with a simple camera control software, but experience shows that it is always preferable to use a much richer and better performing astronomy software – making sure that this one recognizes your camera. The manufacturer's software can be useful for testing and debugging purposes, but be ready to move to another software for your observations.

Once again, we benefit from the huge work that has been done by numerous amateur and professional astronomers: there are several well performing software programs – some even available for free. This is a list of some examples – without any judgment:

- AudeLA[21], free and Open Source;

- Demetra, free, developed by Shelyak Instrument;

- Prism[22], available for purchase, developed by Cyril Cavadore;

- The SkyX, available for purchase in English only;

- AstroArt V6, available for purchase, developed by MSB Software http://www.msb-astroart.com/

- MaximDL[23], available for purchase (in English only).

---

[21] http://www.audela.org/dokuwiki/doku.php?id=fr:start
[22] http://www.prism-astro.com/fr/index.html
[23] http://www.cyanogen.com/maxim_main.php

I don't want to impose here the use of particular software: your choice is a matter of personal taste and adaptation to your instruments (some software can control the whole observing system). I personally have been using AudeLA for a long time, and it suits my needs well- all images in this book are acquired with this software.

When you choose your software, verify that it allows at least the following operations:

- acquisition of simple images or a series;

- continuous acquisition (for the tuning phase);

- control of the exposure time (including 0 sec exposure for bias frames);

- fine tuning of the visualization thresholds;

- recording of the images in FITS format;

- freedom of renaming the saved files;

- binning capabilities (i.e. read out of pixel groups);

- read out of intensity and position of each pixel;

- statistical measurements on image areas;

- window acquisitions (only part of the image);

- camera temperature control;

- image magnification (zoom);

- measurement of the full width half maximum (FWHM) of a star along the X and Y directions;

- precise measurement of the star centroid;

- crop images;

- simple image operations (constant flux addition, image stacking, etc.).

The FITS Format

I strongly recommend saving all your astronomical images in FITS format. It is *the* standard in astronomy. This very old format has a header where we can list all the parameters of the image (date and location of the acquisition, instrument used, etc.). The data is stored with no alteration (no compression), contrary to most formats commonly adopted in photography and printing (jpeg, for example), which severely compress the data at the price of information loss that is unacceptable for scientific purposes.

Later, we will look in more detail at the content of the FITS header.

An important detail: the FITS format can store both the raw image (2D produced by the camera) and the 1D spectral profile (the curves obtained after the data reduction). Therefore, using this format, you can use all the astronomical computing tools, both in imaging or spectroscopy.

7.4 Some Simple Manipulations

Take the time to be familiar with your camera and acquisition software. I will give you some practical exercises that will improve your speed later. These exercises should be done without any lens or telescope, but with the detector in darkness (total or relative, depending on the exercise) – the easiest way is to use the camera cap.

Take a Bias Image

A bias image must be taken in total darkness with a zero exposure time (fig. 7.4). If the camera were perfect, we would get a perfectly dark and uniform image. Is this what happens?

Measure the intensity of some pixels in the image. In most software, just move the mouse over the image, and the information about the pixel below

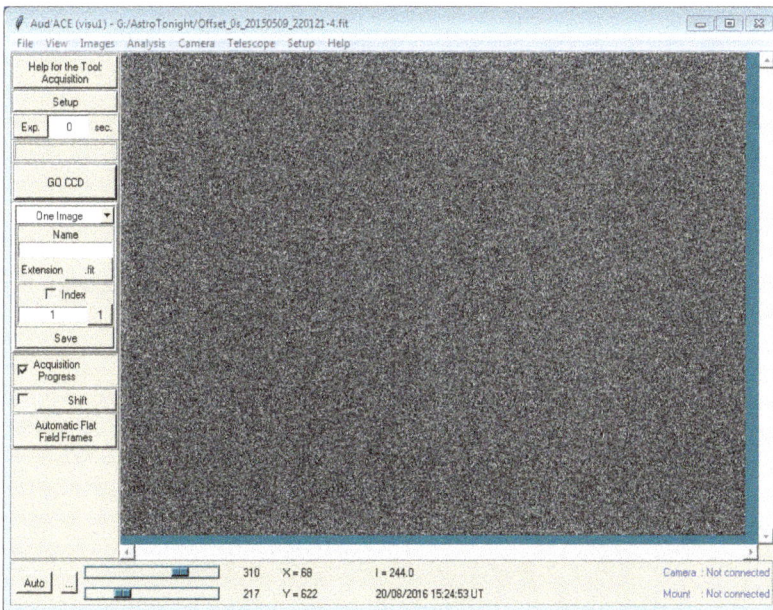

FIG. 7.4 – Raw bias image.

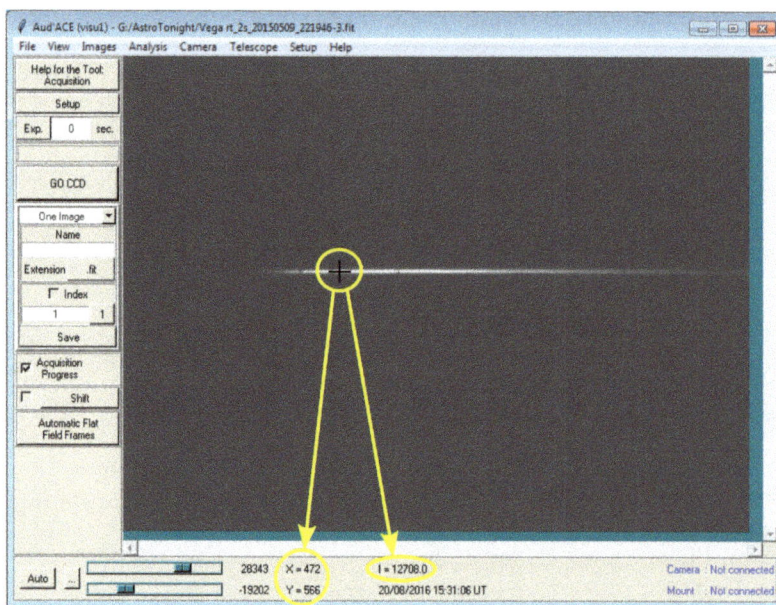

FIG. 7.5 – Intensity of a pixel hovering the mouse cursor.

the cursor (coordinates and intensity) are dynamically shown on the screen (fig. 7.5).

Not only is it that the intensities are not zero, but you can see differences in the measured values from pixel to pixel. You can also measure these values for an ensemble of pixels. In AudeLA, you can select a region in the image, right click, and chose "Statistics in the frame" (fig. 7.6).

This function measures the average level, the median, and the standard deviation in the selected region.

Very often, only a portion of the image is shown on the screen. It is then useful to look at the image full screen (each pixel of the screen then covers multiple pixels of the image). To select precisely a given pixel, it is also useful to look at the image on a 1:1 scale, i.e. a pixel on the screen represents exactly a pixel on the image. Finally, in certain cases, it is useful to zoom in the image – which leads to the appearance of image pixelization. Do these operations with your software.

Measure the Size of the Image

This is a very simple operation, useful to check that the detector of your camera is exactly what you thought. We have seen that moving your mouse on the image shows the coordinates of the pixel below the cursor. Thus, it is sufficient to put your cursor on the four corners of the image and find the one

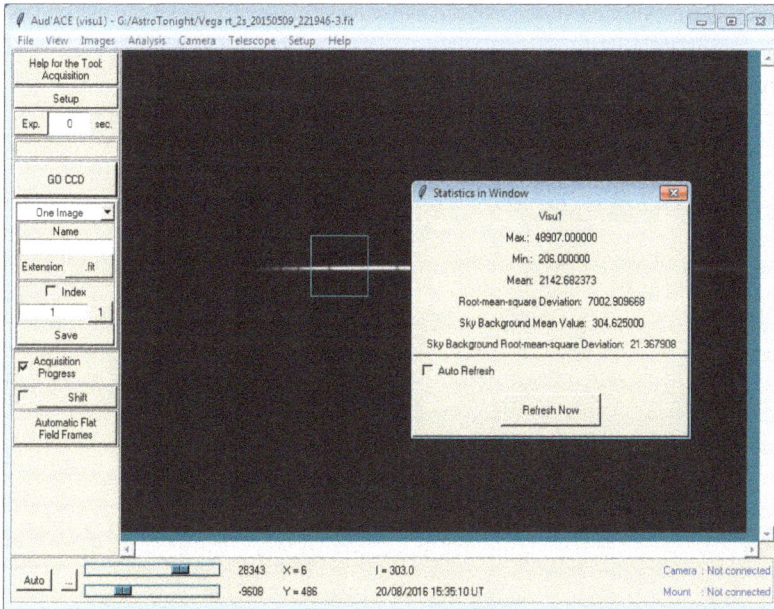

FIG. 7.6 – Statistics on an image sub-region.

corresponding to the larger values – with AudeLA this is the one in the top right corner. The coordinates of that pixel give the total number of pixels in each line and column of the image.

You can now take a second bias image, but now bin it 2×2, i.e. group the pixels two by two at read out. Four pixels now are read out as one pixel. What is the new size of the image? It has been divided by two in each direction, and is now just one fourth of the size of the original image.

Measure the Dynamic Range of the Image

As we have seen, the read out dynamic range of the image is the maximum intensity level the camera can measure for each pixel before saturating. It is a technical specification given by the camera manufacturer, but it is also easy to recover it. Just open the camera cover, and take an exposure of a few seconds with the ambient light. Inevitably, this will saturate your camera pixels, and all the pixels in the image will be at their maximum possible level, determined by the camera read out dynamic range. On this image, you will see that all pixels have the same large value. Because the electronics that does the conversion works in binary, as all electronics do, the large value will always be a power of two, or, to be more precise, a power of two... minus one, to include the value zero.

A simple example: with an Atik 314L+ camera, the maximum value is 65,536 ADU. To infer the read out dynamic range of the converter, it is sufficient to add one (obtaining in this case 65,536), and find to which power of two this new value corresponds. In this specific case, the camera read out dynamics is 16 bits, since $2^{16} = 65,536$. Recent cameras have dynamics generally in between 12 and 16 bits, but you can also find some cheap cameras (possibly color-sensitive) with "only" 8 bits, i.e. with 256 gray levels.

Play with the Visualization Thresholds

It is probable that the last (saturated) image shows up completely white on your screen. No matter what you do with the visualization threshold for your computer screen, it stays white.

Take another bias image (see previous paragraph). Now your image is probably showing up as "mottled" with pixels of different gray levels.

Find the visualization thresholds – in common software they are chosen automatically based on the range of levels in the image.

Look at the image in its full dynamic range, i.e. with the lower threshold set to zero, and the upper threshold to the maximum possible value (65,536 ADU for the Atik 314L+ camera). The image is totally black now, isn't it?

For a bias image, even if each pixel has a non-zero value, these values are much smaller than the maximum dynamic range of the camera (a few hundreds ADUs in my case). Consequently, when we show the image with the full dynamic range, all the pixels are dark.

Now set a very low upper threshold: keep the lower one at zero, and set the upper to 100 ADU for example. What do you see? The image is white! This is not surprising, since all the pixels have a value larger than 100 ADU; and therefore they are shown as white with these thresholds.

To recover the initial display, you can use the function that your software provides to automatically reset the thresholds based on the dynamic range of the image. But you can also measure the mean intensity and the standard deviation on the image and chose the thresholds based on these values (for example, mean value minus standard deviation for the lower threshold, and mean value plus standard deviation for the upper threshold).

These manipulations demonstrate that the same image can look very different depending on the visualization thresholds adopted.

Evaluate the Camera Sensitivity

This is a recreational exercise, to "see" the sensitivity of your camera. So far, you have taken a bias image (in darkness), then a saturated image (opening the cover). Now try to take an image in between these two extremes, i.e. neither black, nor saturated. To do so, try several methods:

- Put the cover on the camera, and start an exposure of a few seconds. During the exposure, quickly remove and replace the cover. Can you be fast enough so that the camera does not saturate?

– Remove the cover, and put it on a table. Put the camera, with the detector oriented towards the table, on it. Start a very short exposition, say 0.1 s. Do you get a signal? If not, try again lifting a bit the camera, and increase the height until you get a level high enough.

– Turn the camera sensor toward the ceiling, and put a white piece of paper over the front of the camera. Take a short exposition for 0.1 s. Is the image saturated? Probably yes...in this case, put another piece of paper, then a third, etc. until an exposition of 0.1 s does not saturate the camera anymore.

The goal of this exercise is to demonstrate the extreme sensitivity of the CCD camera. The fact that soon you will be taking exposures of several minutes from behind a telescope (and you will be far from saturation) makes one understand how faint a star really is.

Measure the Read Out Noise

This is a slightly more technical exercise, but it has spectacular outcomes too. Take two successive bias images and save them – in FITS format, of course. Since you are taking the two images in the same exact conditions, each pixel should have the exact same value in both of them. From the previous exercises you can expect that this will most likely not happen. You can try to measure the position and ADU level of a bright pixel in your first image, and then compare it to the value of the same pixel in the second image.

In my case (Atik 314L+), I have 336 ADU in one image and 323 ADU in the other. What causes this difference (13 ADU)? It is the read out error (or read out noise) of the converter that I already mentioned. To have a statistical measurment of this noise, you can take 5, 10 or 100 images, repeat the exercise and take the average of the measured differences.

However, there is a smarter way: the image contains thousands of pixels, i.e. for each image you take thousands of readings! Of course, each pixel has its own sensitivity, but taking the difference between the two images (since the sensitivity of the pixels is the same in both), we are left with the difference of the measurement only (more precisely, the read out noise is counted twice in each pixel).

Let me make an important remark: taking the difference between two images means your software measures the intensity of each pixel in the first image, and subtracts from it the intensity in the corresponding pixel of the second image. The resulting image is thus the outcome of the difference between the two original images pixel-by-pixel.

To avoid measuring each individual pixel – which would be tedious – it is sufficient to select a (large) region of this new image, and make a statistical measurement: the mean value should be very close to zero (since we took the difference of two very similar images), and the standard deviation is the read out noise (or, more precisely, the read out noise multipled by a factor of $\sqrt{2}$), i.e. the dispersion in measurements when everything is held constant, except the repetition of readings.

In my case, I find a standard deviation of 6.5 ADU, which corresponds to a read out noise of 4.6 ADU (this value is in perfect agreement with the specifications of my camera).

Compare the read out noise (4.2 ADU in my case) to the camera read out dynamic range (65,536 ADU). We do measure a certain dispersion, but it is tiny compared to the camera capabilities.

Reduce the Read Out Noise

In the previous exercise, you have measured the read out noise of your camera. I now suggest a simple exercise showing how to reduce such noise.

Take two series of 16 bias images, and then take the median of each series. Use the two resulting images for a new measurement of the read out noise. It should have decreased roughly by a factor of 4 (do you agree?). This is a statistical effect that is very effective in the case of random errors (I will come back to this later), and the read out noise is reduced by the square-root of the number of images (in this case $\sqrt{16} = 4$).

Look at the FITS Header

Use your acquisition software to read the header of one of the FITS files generated in the previous exercise. You should see something like in figure 7.7.

Each line has a keyword, a value, a unit, and a comment.

Some keywords are mandatory by definition of this format. For example, the keyword NAXIS gives the number of axes in the data. Here NAXIS = 2, i.e. the file contains data on two axes. This is trivial: it's an image!

The keyword NAXIS1 gives the dimension of the image along the first axis: it is the image length in pixels that we measured previously by moving the mouse in the image corners. Of course, NAXIS2 corresponds to the image height.

Other keywords are proper to the software or are defined by convention. For instance, the keyword EXPTIME gives the exposure time. It is often useful to have this information saved automatically in the file, since it is almost the only way to retrieve it a few days after the observation.

I will let you find out the meaning of the other keywords. You can find the details of the common conventional keywords on the NASA website[24]. But there might be keywords specific to your software: the FITS standard is very permissive in this sense.

---

[24] http://fits.gsfc.nasa.gov/fits_standard.html

```
En-tête FITS (visu1) - //VBOXSVR/windows/AstroTonight/Vega rt_2s_20150509_221946-3.fit        ⊠

  BGMEAN = 271.75  mean value for background pixels      adu
  BGSIGMA = 17.372866397489531  std sigma value for background pixels     adu
     BIN1 = 1
     BIN2 = 1
   BITPIX = 32
   BSCALE = 1  linear factor in scaling equation
    BZERO = 0  zero point in scaling equation
   CDELT1 = 1  Wavelength pixel step (Angstrom)
  COMMENT =    = 'Valbonne'             / descriptive comment
 CONTRAST = -3.710614200000000e+007  Pixel contrast     adu
  DATAMAX = 48907  maximum value for all pixels     adu
  DATAMIN = 159  minimum value for all pixels     adu
 DATE-END = 2015-05-09T22:20:00  Date of observation end
 DATE-OBS = 2015-05-09T22:19:58  Date of observation start
  DETNAME = Atik 314L+
  EXPTIME = 2  [s] Total time of exposure     s
   EXTEND = T  Extensions are permitted
 INSTRUME = Alpy 600
     MEAN = 371.03827921692653  mean value for all pixels     adu
  MIPS-HI = 445.478668213  High cut for visualisation for MiPS      adu
  MIPS-LO = 167.512802124  Low cut for visualisation for MiPS     adu
    NAXIS = 2  Dimensionality
   NAXIS1 = 1391
   NAXIS2 = 1039
  OBJNAME = Vega
 OBSERVER = F. Cochard
    SIGMA = 1436.8843947249659  std sigma value for all pixels      adu
   SIMPLE = T  C# FITS: 09/05/2015 22:20:03
 SITEELEV = 0
  SITELAT = 0
 SITELONG = 0
 SITENAME = Valbonne
 SWCREATE = SideReal Light v2.0.0.8
 TELESCOP = C8
      TT1 = IMA/SERIES STAT  TT History
```

FIG. 7.7 – FITS header.

Optimize the Exposure Time

A last exercise is to finish our familiarization with the camera. Go back to your camera configuration with the detector pointed at the ceiling and a few white sheets of paper to avoid saturation. I now ask you to choose the exposure time so that the average ADU level of the image is 80% of the read out dynamic range of the camera.

To do so, you have to measure the average level in the initial conditions. You need to measure the target level (in my case, 80% of 65,536, i.e. 52,428 ADU). Now you just need to rescale the exposure time using these two values.

Supplement

After doing these exercises on the acquisition camera, you can start over with the guiding camera, especially if you use different software for it – one of the aims of these exercises is to make you familiar with your software tools.

The next step (I will let you do it yourself) is to do the same kind of operations, but now with a lens in front of the camera – either a camera lens or telescope. You will not work with light randomly hitting the detector, but with an image formed on its surface.

Chapter 8

Adjusting the Spectroscope on a Table

If you only remember one thing from this book, it should be this : start working with your spectroscope on a table, during daylight, independently from the telescope. Put the telescope and the spectroscope together only when you master both. There are many reasons for this:

- the two instruments are very independent and work perfectly well on their own;

- almost all adjustments on the spectroscope are easier to make during daylight (and some are almost impossible to do by night!);

- each element has its own complexity, which you need to master. Wanting to learn everything at once is a risky bet;

- it is much easier to work comfortably sitting in a warm room with a table, rather than having to bend around your telescope in the dark;

- there are many playful and educational experiments to be done with different light sources before tackling stars.

In this chapter, we will start to work with the spectroscope, and make it produce an image of the spectrum of the light entering its slit. We will also see how to deal with guiding at the end of the chapter.

The advice is generic, i.e. it is not bound to any specific spectroscope. However, since I want to show real images of what I discuss, I will be using the Alpy 600 spectroscope to illustrate the procedures below.

8.1 Which Light Source

We have several easily accessible light sources to make these first steps. Here are some of them:

– The most obvious: the Sun! All the light during the day, except artificial illumination, comes from the Sun. No matter if it is sunny or it rains, or if there is a blue sky or clouds, such light is perfectly identifiable at first glance in a spectroscope. Obviously, depending on the weather, on whether you observe the sunlight directly, or by reflection, the spectrum will be modified, but you can always identify the very specific solar spectrum (fig. 8.1).

– An energy saving lamp, such as the common lamps you use in your everyday life. These lamps have very interesting spectra (fig. 8.2).

FIG. 8.1 – Raw image: 2D spectrum of the Sun.

FIG. 8.2 – Raw image: 2D spectrum of an energy saving lamp.

– An incandescent lamp, which is now more difficult to find – because of the energy saving policies – but still quite easily available. Such a lamp has a continuous spectrum, with depends only on its temperature (Blackbody spectrum – fig. 8.3).

FIG. 8.3 – Raw image: 2D spectrum of an incandescent lamp.

– A flame, very similar to the previous one (Blackbody spectrum) (fig. 8.4).

– A white LED (Light Emitting Diode, fig. 8.5), which shows a large bump in the blue, and smaller ones in the green and red. Note that you can find LEDs of different colors.

FIG. 8.4 – Raw image: 2D spectrum of a flame.

FIG. 8.5 – Raw image: 2D spectrum of a white LED.

– Calibration lamp. It is a lamp which contains a particular gas (neon, argon, mercury, thorium, ...) and thus produces the emission spectrum characteristic of that gas. Some laboratory lamps are extremely expensive, but there are also very affordable calibration lamps. For example, the night lights for children (available for a few dollars in any supermarket) have small neon bulbs in them (fig. 8.6).

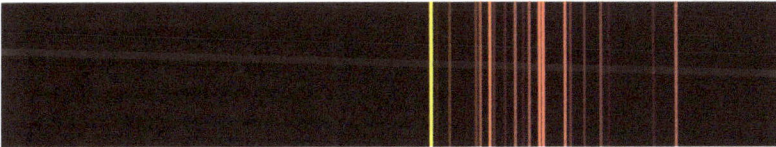

FIG. 8.6 – Raw image: 2D spectrum of neon.

– You can also study the spectra of the different lamps around you.

You can also do other kind of observations: compare a light source observed directly and then with a colored filter. Use a piece of colored paper illuminated by white light, etc. The pool of choices is large, even without spending a single dollar. I promise that even these first observations will surprise you and raise questions.

When the spectroscope is mounted on the telescope, the light from stars is focused in a cone determined by the telescope focal ratio (the well known F/D ratio), i.e. it is concentrated toward the slit. Conversely, when the spectroscope is in front of you on a table, all the ambient light can enter the slit if you don't take any precautions. It is very likely that the light will enter with a focal ratio much larger than what you would get with a telescope. No big deal – at least at the beginning – since the light arriving at a large angle will not get through the collimator (vignetting effect). On the other hand, you want to avoid having a point-like source in front of the slit: this can modify the spectrum. To avoid these effects, work with diffuse and more or less homogeneous light in front of the slit, for example using a piece of translucent paper at the instrument entry, or a sheet of paper, like in figure 8.7.

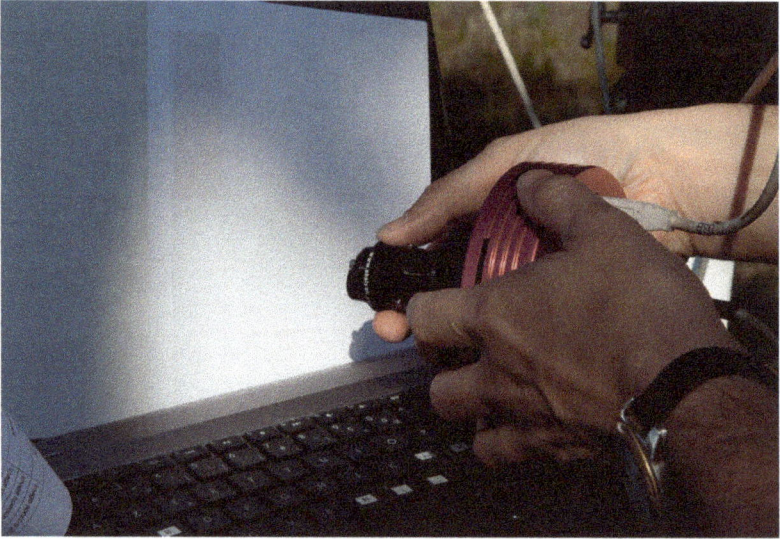

FIG. 8.7 – Using something to diffuse the light at the spectroscope entry.

Finally, depending on your instrument, it might be preferable to carry out the set-up using a calibration lamp (for example, a neon lamp) or with daylight. Consult your documentation to make the proper choice for your instrument. In the former case, you will have an emission line spectrum, while in the latter you will have a continuous spectrum with a multitude of absorption lines. In the example I use here (Alpy 600), the recommendation is to use sunlight.

8.2 Install the Acquisition Camera

Install your acquisition camera on the spectroscope. To do this, you need to fulfill two mandatory conditions:

- the CCD detector needs to be placed at the right distance from the spectroscope;

- the assembly needs to be very stiff, to prevent any differential motion when the telescope moves to point in different directions.

Nowadays, most cameras have a standard interface (fig. 8.8) – $M42$ thread with a pitch of 0.75 mm, which is also known as *T-mount*.

FIG. 8.8 – Camera installation with a T mount.

Make sure you have the right adapter rings. Tighten up the ring and screws strongly (but not too tight!) to ensure the best stiffness. You need to be able to push on the camera without detecting any movement.

If the camera is loose it will move when the telescope turns. Consequently, the spectrum will shift on the CCD. This makes a proper calibration almost impossible.

8.3 Focusing and Orientation

Here we deal with fine tuning phase of the spectroscope setup: the focusing, and then the orientation of the acquisition camera to the spectroscope focus. The first trials are usually the hardest:

– the spectroscope is usually far from its optimal settings;

– one doesn't really know what image to expect;

– one doesn't know, a priori the amount of the adjustments to be made.

To begin with, have a look in the spectroscope documentation for the type of image you aim for with the recommended light source for the setup (Sun or calibration lamp). With the Alpy 600, you aim for the image in figure 8.9 (note that this image was obtained under direct sunlight, and you can spot a few light leaks at the edges).

FIG. 8.9 – Nominal image of the Sun spectra with an Alpy 600.

Take a first image with a short exposure time – for example 0.1 seconds with an Alpy 600. Adapt the visualization thresholds to clearly show the image. You should get something like in figure 8.10.

Let me remind you that it is better to have a homogeneous light source (which yields a cleaner spectrum). Use some translucent paper, or a white piece of paper at the spectroscope entry (see fig. 8.7). Make sure your image is not saturated. If it is, adjust the exposure time to correct it. When you have found a seemingly appropriate exposure time, start continuous acquisitions, and modify little by little the focusing. When you modify the focus, the spectrum moves on the image because the mechanical parts wiggle a little bit (it might even rotate, depending on the spectroscope). Often these motions happen during the acquisition itself, producing perturbed images (fig. 8.11).

To judge the improvements or degradation caused by the new settings, you need to wait for the second image without touching the instrument – if

FIG. 8.10 – First image of the solar spectrum with the Alpy – exposure time of 0.05 seconds.

you continuously play with the settings without waiting, there is no way to know if you are going in the right direction.

The closer you get to optimal focusing, the finer the details you can see in the spectrum. If you use sunlight to focus (as for the Alpy), choose a not too deep absorption line at the spectrum center, and look for the best settings to get the highest contrast for that line. If you use a calibration lamp (as with the LISA and Lhires III), the emission lines must be as thin as possible; you can measure their width (full width half maximum, FWHM) with your acquisition software.

Find the best possible setup; and to make sure it is the optimal setup, overshoot it and then come back to it. You might need to adjust the exposure time: when the spectrum is focused the details are finer, and the light is less dispersed which might cause saturation (if this happens, adjust the exposure time).

Then, set up your camera so that the spectrum is horizontally aligned. Depending on the instrument, this has to be done before or after focusing. The spectrum inclination can (and should) be less than 1 degree. The inclination will be corrected during data reduction (depending on the software tools you use), but it is always better if these corrections are minimal.

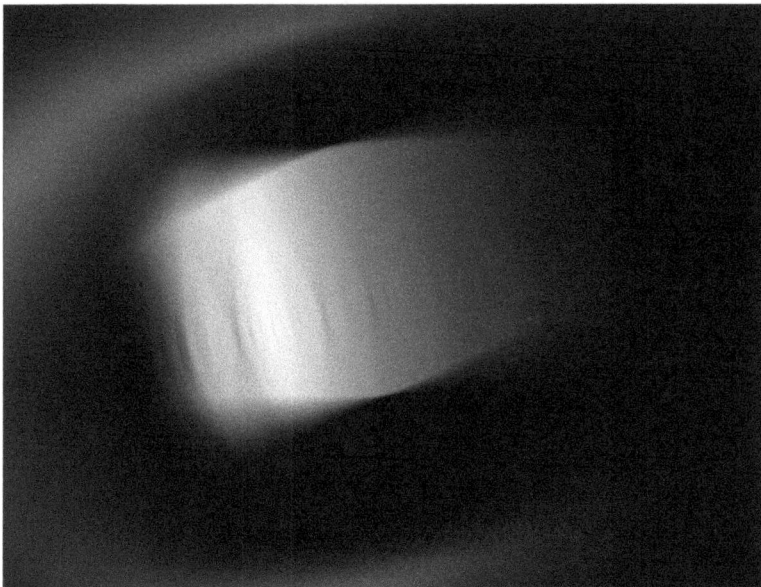

FIG. 8.11 – Raw image during the setup process.

8.4 Blue on Left, the Red on the Right

The image showing on the screen needs to comply with the astronomical convention that blue (high frequency and short wavelength) needs to be on the left, and red (low frequency and long wavelength) on the right.

I can see you frowning your eyebrows: how am I supposed to know where the blue and red are, since I have a black and white camera? This is a delicate point at the beginning, since you need to recognize a few characteristic lines in the spectrum. Use your spectroscope documentation – in the case of the solar spectrum taken with the Alpy, I can easily recognize the H&K lines of calcium, which are in the deep blue and which therefore should be on the left of the spectrum (fig. 8.12).

If you have perfectly set the focus and orientation, and realize just now that the image is in the wrong direction, you can take advantage of a trick available on most cameras: mirror the image horizontally (which inverts the left and right sides of the image). This can usually be done with the camera parameters. However, be careful: if you invert the spectrum image, you will have to invert all the images you take while observing (dark, bias, flats, calibration...) to have a consistent data reduction. Make sure the mirroring parameter is saved when you quit the acquisition software.

When you have a well focused, horizontal, and properly oriented spectrum on your screen, you have completed the main steps in setting up your

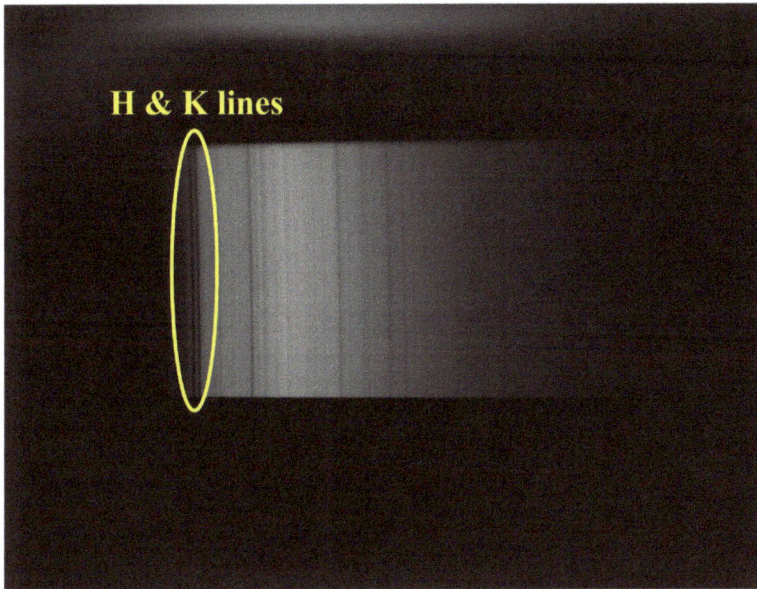

FIG. 8.12 – Calcium H&K lines in the solar spectrum of an Alpy.

spectroscope. All the rest are just operations to bring the light of stars into your properly set up instrument – and to properly use the resulting spectrum.

8.5 Choice of the Range of Wavelengths

This step does not always apply and depends on which spectroscope you are using. If it is a low-resolution instrument (such as the Alpy 600), it is likely that you cannot change the spectral domain. But if it is a medium or high-resolution spectroscope, it is likely that you can change the angle of the diffraction grating, and thus the observed spectral domain.

This can be a very complicated operation for beginners: once again you need to recognize the spectrum. To begin with, I strongly recommend that you work around the H_α line at 656.3 nm. In fact, it is definitely the easiest line to recognize in almost all stars, including the Sun. It is a deep absorption line in the deep red – if you need, and if your instrument allows for it, take out the camera, and look at the spectrum through an eyepiece (your eye has the advantage of seeing the colors).

With some practice, it is a setup that only takes a few minutes. But the first time, it can take several hours to find some fixed marks and understand what is seen. Take your time: it is not wasted! Find a database of lines corresponding to the resolution of your instrument, and learn how to "move" along the solar spectrum.

When you are sure you have found the H_α line (or any other you might be looking for), record the grating position. This way, if you lose the setup during the observations, you can get back to it easily.

8.6 Setting up the Guiding Camera

The spectroscope is now set up and operational. Before moving on, take the time to also set up the guiding system. Until now we used an extended and generous light source (Sun or calibration lamp). When you observe a star, it is a point-like source which needs to be properly aligned on the spectroscope slit.

While the spectroscope is on your table, and it is not mounted on the telescope (or any other optics), no image is formed on the slit plane. On the other hand, it is easy to see the slit itself as a thin dark line. Setting up the guiding camera consists in focusing to get the best such line (i.e. make it as thin as possible, a few pixels maximum), and orient it horizontally (or vertically in a few cases) in the guiding image (fig. 8.13).

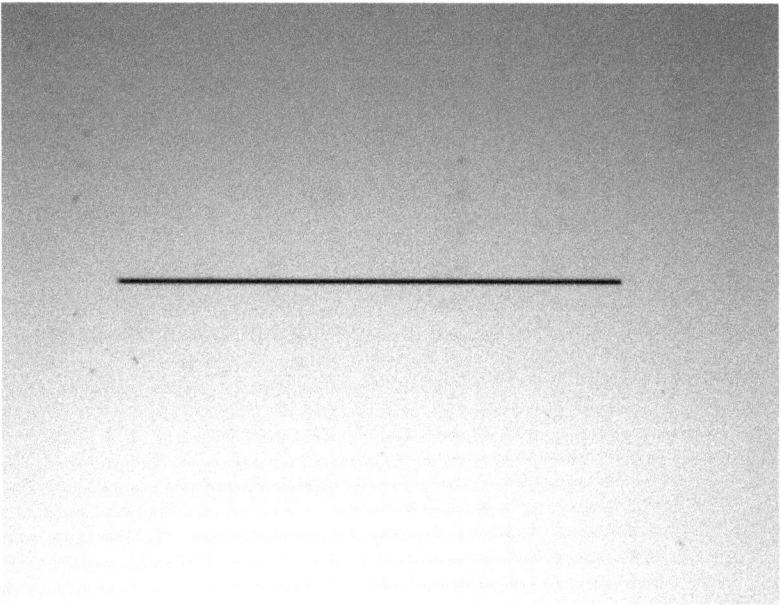

FIG. 8.13 – Slit in the guiding image.

The orientation of the slit in the guiding image does not influence the orientation of the spectrum in the main image. But we will see later that if the slit is well aligned horizontally, movements of the telescope are easier to understand and control.

Chapter 9

Physical Measurements and Data Reduction

At this stage, you're ready to take your first spectra on the table: the spectroscope is ready to take the first raw images. I believe it is useful to have a detailed look at how to use those images, since they are not the goal themselves, instead they are the way to get a 1D spectral profile. The whole art of spectroscopy is in making a *physical measurement* of the observed object. The scientific information we are looking for is embedded in the 2D raw image, together with a plethora of many instrumental effects.

In some simple cases, the 2D raw image is sufficient, for instance, if you only want to convince yourself that an incandescence bulb and an energy saving lamp do not have the same spectrum. Rarely, however, do raw images suffice for scientific applications. It can happen for instance to show the detection of outbursts in Be stars: it suffices to see whether the H_α line is in emission or absorption – which can usually be seen immediately[25].

Other than these isolated cases, the 2D raw images need to be processed to extract the physical information. The software available nowadays may make these spectra calibration steps automatic (a few seconds are typically enough to complete them), but I encourage you to do the first data reduction by hand, to familiarize yourself with the tools, and better understand how to go from the digital image to the physics of the light source. This is also required to prevent "black box" software from making decisions instead of you.

9.1 Your Mission: the Spectral Profile

Data processing can vary a lot depending on your objective. But in most cases, you need to transform your raw data into a 1D spectral distribution (fig. 9.1).

---

[25] In reality, it is not always so obvious: if the line is weak in emission, it can be hard to see it in a raw spectrum.

HD119850 - C8 Alpy 600 - 18 Apr 2015 - 3x300s

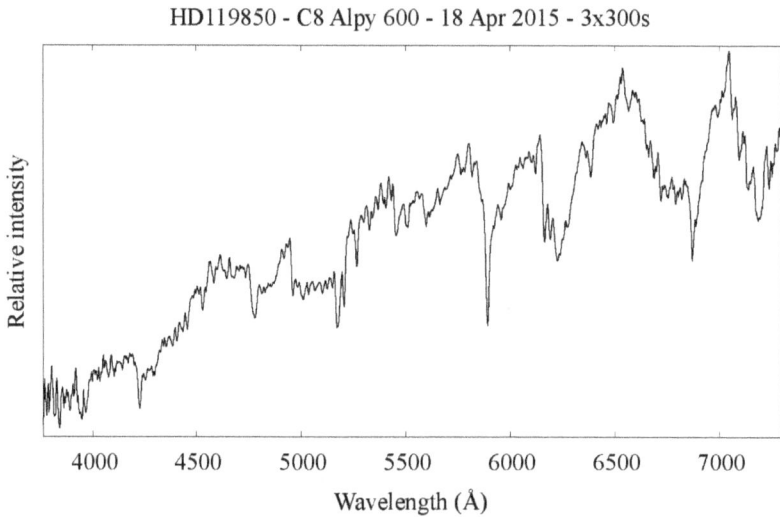

FIG. 9.1 – Example of a spectral distribution.

This data processing procedure is called *data reduction* – the same way we reduce a sauce when cooking: you start from a set of ingredients, you follow a recipe, and you obtain an edible result.

You can consider your mission as an observer complete when you have produced a properly calibrated spectrum.

9.2   Do not put the Cart before the Horse

I've seen many amateur astronomers starting in spectroscopy as if it were a stand-alone activity – and not a measurement technique. They consider the acquisition the most important part. It is the gratifying part, certainly, acquired with the telescope. Data reduction is seen as something secondary, which can be delayed, and done only if some spare time is available. Reducing data is often seen as a chore, and many 2D raw images sit on hard drives, that probably will never be converted into the form of an usable spectrum.

I urge you to completely ignore this reasoning, and put the horses in front of the cart: a spectroscopic observation is a complete sequence that starts with an acquisition sequence, and then proceeds to data reduction, which yields to a scientific result.

In reality, the data reduction process is central, and the acquisition should be seen as a necessary step to feed the data reduction process. The images serve the data reduction process, not the other way around.

9.3 Intensity and Wavelength

Spectroscopy provides for a variety of physical data: radial velocities, chemical composition, physical conditions, etc. All these measurements do not come from the same portion of the spectrum. Simplifying extremely, the spectral distribution can be seen as two-dimensional information: one dimension is the wavelength, and the other is the relative intensity. If you want to measure a radial velocity, you mainly care about the horizontal axis (i.e. the wavelength)[26]: therefore wavelength calibration requires close attention. If instead you want to measure the effective temperature of a star, you need to look at the general shape of the spectrum and the instrumental response needs to be accounted for[27]. Finally, if you want to analyze specific lines (their presence or absence, their relative intensities, how they change in time), then the local value of the intensity around those lines is what matters.

It is common to see measurements very carefully calibrated for some of these aspects (wavelengths calibration, correction of the instrumental response, etc.), but much less in the others. When using a spectrum taken by other observers, it is always useful to know the context and aim of that measurement.

Nevertheless, in the following I assume that our spectra need to be properly reduced, in all their dimensions. That is, they can be used for any of their possible scientific applications. The spectra you produce must provide the most accurate information possible about their original sources, without compromising any of their possible future use.

9.4 Distinguishing Technicalities and Science

The ultimate goal of observations is to produce a scientific result. In the realm of amateur astronomy, typically the same person carries out the observation and then uses the data for a scientific use. In the professional world, there is a stronger distinction: rarely is the person who processes the data the person that operated the telescope.

I take advantage of this distinction to underline that the spectral profile is the intersection of the technicalities that produced it and the science that can be understood from it. When you work on your instrument, you wear the technician's hat, which means your aim is to produce an observational spectrum. When you have your 1D spectral profile, and infer the temperature or the velocity of a star, you wear the scientist's hat.

As a technician, you are the only person able to compensate for all the instrumental factors in your measurements : nobody knows your instrument

---

[26] This is oversimplified: high precision radial velocity requires precise profiles, thus good instrumental and atmospherical response

[27] Again, this is oversimplified: a good wavelength calibration is required to identify specific features

better than you. Ideally, the analysis you perform on your spectra should not depend on the observing conditions.

I emphasize this distinction between Technicalities and Science because it has several important consequences :

- it encourages you to take "generic" measurements, without guessing the results beforehand;

- it naturally shows that the process of data reduction is instrument-dependent (while it does not depend on the scientific goal);

- it encourages the observer to always use the same data reduction process, for the sake of reproducibility, quality, and productivity;

- it allows reduction of spectra in a few seconds, in a very automated way, once the reduction process is properly configured and set up (also called a "data analysis pipeline");

- even if you keep your spectra for yourself (what a waste, though!), it will be easy to go back to them a few months or years later, when you will have forgotten everything about the observing conditions;

- and... I am convinced that you will quickly get involved in collective observational programs, or even collaborations with professionals: in this case, you necessarily have to produce generic-purpose spectra. So, develop good habits now!

9.5 Systematic and Random Errors

A spectroscopic observation is first of all a physical measurement, which yields the relative intensity as a function of the wavelength of the light emitted by an object. Let me recall here a few obvious aspects of this kind of measurement.

- Any physical measurement is necessarily characterized by some degree of *uncertainty*, or *error*. Depending on the nature of the uncertainty, this can be compensated or reduced.

- Rigorously, one should never provide a measurement without quantifying its uncertainty. Too often this rigor is not respected; nevertheless, one should always worry about whether the instrument can achieve the sensitivity required for a measurement, and verify that the outcomes are consistent.

- The errors can be of two kinds: *systematic* and *random* errors.

- If you measure a length with a ruler, it is unlikely that they will be identical, since the ruler itself is manufactured with a limited precision: you make a systematic error, since it is the same for all measurements.

- If you ask several people to measure the same length with the same ruler up to one tenth of a mm, it is likely that all measurements will be different. This is because of the differences in the ruler positioning, the interpretation of each person, etc. It is an example of random error.

- The randomness of an error can also arise from a variation in the observed phenomenon itself.

– A systematic error can be compensated with a calibration process consisting of comparing our measurements with a standard reference assumed to be perfect (however, the calibration has an uncertainty itself).

– A random error cannot be completely eliminated, but it can be mitigated by repeating the measurement a large number of times, and then taking the mean (or median) of the values obtained. The greater the number of measurements, the smaller the random error. If each measurement is independant it decreases as the square root of the number of measurements: if you make the same measurement 100 times, you reduce the random error by a factor of ten, since $\sqrt{100} = 10$.

– You can always estimate the random error on your measurements by replicating them, and measuring the variation of the obtained results. The average value of these results then becomes the best estimate of the real mean value, and the typical variation (i.e. the mean of the distances from the best estimate) is the uncertainty (or, to use a more modern vocabulary, the precision of the measurement).

In spectroscopy, both kinds of error are always present. Because of this, it is common to rely on sets of images rather than single images, during calibration processes.

9.6 Signal-to-Noise Ratio

You will constantly hear people talking about the signal to noise ratio (SNR). It is a parameter describing the ability of our instrument to yield a reliable value for the quantities we want to measure.

The signal is the information we want to measure. For example, the intensity of a spectral line. The noise corresponds to the measurement uncertainty – the name comes from the acoustics domain, in which it has a natural interpretation: the noise is what covers the details in the sound.

When the signal is high compared to the noise, we can trust our measurement – the SNR is important. If somebody talks loudly in a quiet room, you can perfectly understand what they say. Conversely, if you try to listen to somebody talking with a low voice in a noisy environment, you will have to ask the person to repeat what they said to make sure you got it – in this second case the SNR is low.

This description of the SNR is qualitative. We can transform this into a mathematical ratio between the signal and the noise, as long as they are expressed in the same units (the SNR is a dimensionless quantity). For example, if I measure a line intensity of 450 ADU with a noise level of 3 ADU, the SNR is:

$$\text{SNR} = \frac{\text{Intensity of the line}}{\text{Noise level}} = \frac{450}{3} = 150$$

In practice, it is common to have SNR values larger than 100 for very deep lines, while caution is required when looking at claimed results with SNR lower than 10. However, SNR in this range are common in astrophysics, since we observe very faint objects and often at the detection threshold for the instrument.

9.7 The Steps to Reduce the Data

Whether you reduce your data manually, or use automated software, you always follow more or less the same steps:

- pre-processing;

- geometric corrections;

- spectrum extraction;

- wavelength calibration;

- re-sampling and linearization;

- compensation of instrumental and atmospheric effects;

- scaling;

- exporting in a standard format.

Pre-processing

The purpose of pre-processing is to compensate for the intrinsic defects of the camera (bias, dark current, read-out noise, non-uniform pixel response, hot pixels) and some optical defects of the instrument (vignetting, presence of dust). For this, we use a series of bias, dark, and flat images.

Geometric Corrections

On the raw image, you can see that, despite all your efforts, the spectrum is not perfectly horizontal. Moreover, it is likely that your spectroscope will slightly deform the spectrum (inclination or rounding). The *geometric corrections* aim to align the spectrum on the two normal axes (vertical and horizontal). Among these corrections, there is at least one rotation of the image (so-called *tilt*), to which, in some cases, we add an inclination correction (so-called *slant*) and a shape correction (so called *smile* in some software like ISIS). These corrections can vary or be complemented by other operations, but their aim is always the same: to align the spectrum on the image axes for extraction.

> 💡 Even if the imperfect horizontal alignment is compensated during the data reduction, you need to reduce it as much as possible during the instrument setup: the less the raw data are manipulated, the less we alter their scientific content.

Note that the inclination of the spectrum on the image can create artifacts, i.e. a "staircase step profile". The narrower the spectrum, the more important this effect is – this is yet another reason to take care of the inclination during the instrument setup process.

Profile Extraction

The extraction of the spectral distribution is the central operation in data reduction: we want to transform our 2D image into a 1D spectral distribution. Although this is a seemingly simple operation, the spectrum is spread over multiple pixels. Thus, first we need to identify which pixels must be considered, and sum them in each column.

For stellar spectra taken with a slit spectroscope, we have on each side the sky background spectrum consisting of intrinsic atmospheric emission lines and continuum and emission lines from light pollution (fig. 9.2). In the extraction operation, this background spectrum which is also behind the stellar spectrum can be removed, leaving only the spectrum of the target star.

For spectra of extended objects, we do not always have this information on the sky background, therefore it is impossible to apply this correction.

Wavelength Calibration

The extraction operation gives an energy profile associated to each column of pixels at right angles to the stellar spectrum. To use this information, we need to associate each column to its corresponding wavelength expressed in physical units (nm or Å): this is what the *wavelength calibration* does.

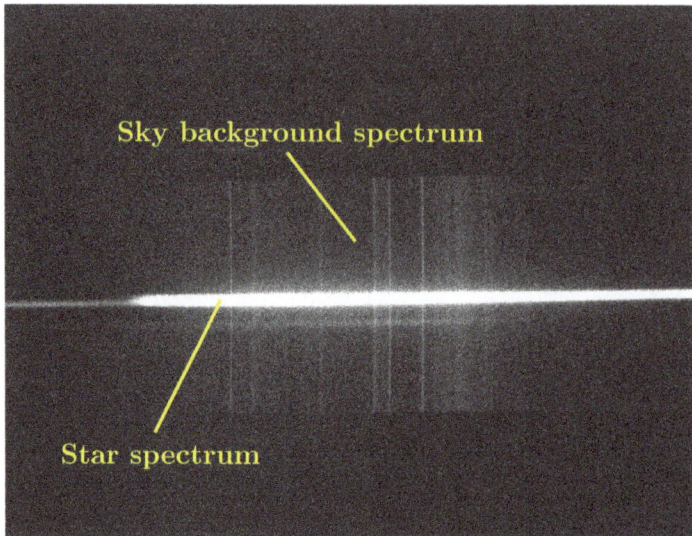

FIG. 9.2 – Image of a stellar spectrum with the sky background. The visualization thresholds are adjusted to enhance the spectrum of the polluted sky (image taken in suburban outskirts with exposure time of 60 s).

Since the formation of the spectrum follows purely deterministic rules, which are used to design the instrument, so one might think that the relationship between the pixel column position and the corresponding wavelength is given by the spectroscope manufacturer. However, the manufacturing tolerance for the mechanical and optical components does not allow for a calculation precise enough for scientific purposes. In practice, you need to calculate the relationship between pixel position and wavelength – the so-called dispersion law – for each setup, and each time you change anything in your instrument (remove it from the telescope, remove the camera, change the grating angle...).

In reality, there is a very efficient approach which allows you to bypass all the manufacturing uncertainties: it consists in taking a known light source, and take a spectrum of it in exactly the same conditions you use for your scientific target. With this known light source, you can identify some lines of known wavelength (in the literature). The data reduction software can then build a *dispersion law* to associate a wavelength to each pixel.

To the first order, we can consider that the dispersion is constant over the whole spectrum – in this case we are using a *linear dispersion law*. This means that the wavelength associated with each pixel is linearly proportional to its position. If this is the case, the dispersion law can be obtained by observing two lines in the spectrum and building a linear relationship of the form:

$$\lambda = a \cdot x + b$$

(where x is the pixel position, and a, b are coefficients inferred from the calibration spectrum.)

Unfortunately, this approach is useful only in very rare cases (only for gratings, not for grisms), and generally yields a mediocre precision of the calibration. In reality, the dispersion in a spectroscope is not precisely linear (it depends on the wavelength), and it is definitely better to use a polynomial relation of the following type:

$$\lambda = a + b \cdot x + c \cdot x^2 + d \cdot x^3 + \cdots$$

(where a, b, c, etc. are coefficients obtained from the calibration spectrum.)

In practice, we generally use polynomials of third or fourth order, and going beyond does not give significant improvements. The more spectral lines you can identify in your calibration spectrum, the better the calibration quality is. You need at least one more line than the number of coefficients you use – 5 to 10 lines is a good order of magnitude for a good calibration (the more lines you take, the better the result). Moreover, to have the best precision in evaluating the polynomial coefficients, these lines must be spread over the whole spectral domain of your profile – it is easy to see (and it can be formally demonstrated) that if we use only lines in a small portion of the spectrum, the extrapolation is not reliable.

When we have more lines than necessary (a polynomial of third order needs at least 4 lines), the dispersion law calculation might result in differences between the theoretical value and the measured value of the wavelength of the extra lines. This information allows you to verify the quality of the dispersion law just established – and more specifically, this allows for a quick check, if we made a mistake in correctly identifying the lines (I am talking from my experience...).

Therefore, a good calibration source contains many lines easy to identify, and regularly spread over the whole spectral domain covered by your spectroscope.

Generally, we use calibration lamps, containing a known gas (neon, argon, xenon, thorium, etc.) excited by high voltage input (about 100 to 300 volt). These lamps have the advantage of providing a spectrum of emission lines of known wavelength, which is exactly what we need.

Neon lamps have some peculiarities particularly interesting for us. Beware, the neon calibration lamps should not be confused with "neon tubes" common for indoors illumination. The latter are called this for historical reasons, but have not contained neon for a long time.

The real neon lamps have important advantages:

– They are very common for every-day-life applications (pilot light, children night lights...), and this makes them cheap;

- They require a moderate voltage to be excited (of the order of 80 volts) and thus can be used by plugging them directly into the wall (120 volts), without the need for a high voltage adapter;

- they have many useful lines around H_α, which is the main line in stellar spectra.

These properties make neon lamps the preferred choice of amateur spectroscopists. Nevertheless, the miracle is not complete, because neon lamps have a spectrum mainly limited to the yellow and red (which explains the yellowish color of this light), and if you work in another wavelength domain or at low resolution, neon does not suffice.

Nowadays, we can find lamps with a mixture of neon and argon to compensate for this problem – since argon has a lot of blue lines. However, these lamps are slightly harder to use (input voltage required of 300 volts).

On top of the calibration lamps, there is another kind of source astronomers can use: the stars themselves. Since they are hot, they solve the problem in the blue range. Hot stars (spectral type A or B) are particularly suitable for wavelength calibration, since they show Balmer lines (hydrogen lines)[28]. But be very careful when selecting such a star.

In some cases, it is possible to combine the two sources – calibration lamp and hot star –, to ensure a precise calibration over the entire spectral domain (roughly speaking, the lamp covers the red portion and the star the blue portion). This is not ideal, of course, but can help if you don't have a good calibration lamp.

Re-sampling and Linearization

If we stop at applying the dispersion law to the raw profile, we know the wavelength of each pixel, but the change in wavelength from pixel to pixel remains variable and unknown. In theory, this is not a problem, but in practice it rises issues: when you want to compare two spectra, or when you open your spectrum with different software, the dispersion law must be considered. This would require a standard format to express this law in all software – including non-polynomial forms.

To completely avoid these issues, it is common to *linearize* the spectrum, i.e. re-sample it with a constant wavelength increment. The closer to linear the dispersion law is, the more acceptable this operation is, because it will very slightly alter the data.

Once the dispersion law is linearized, we just need to associate with the profile two values to be able to use it in all cases: the wavelength of the first pixel, and the wavelength increment corresponding to each pixel. No need for other things to compare two spectra anymore!

---

[28] Be careful however: if the star taken as reference has a significant radial velocity, the Doppler effect affects the balmer line positions and hence the dispersion law

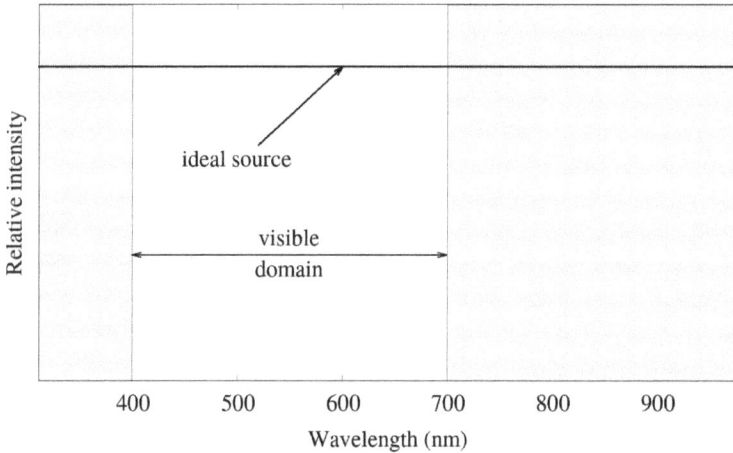

FIG. 9.3 – Spectrum of an ideal white lamp.

In a FITS file, the (linear) dispersion is defined by four parameters:

- CRVAL1 : wavelength of the reference pixel;

- CRPIX1 : reference pixel (generally, the value of this variable is 1, since the reference wavelength is given for the first pixel);

- CDELT1 : dispersion per pixel;

- CUNIT1 : physical units of the dispersion (either in Å or nm).

Correction for Instrumental and Atmospheric Effects

After this operation, the spectrum is calibrated, i.e. the x-axis can be expressed in wavelength. However, it is still strongly affected by the *instrumental response*, i.e. the sensitivity of the instrument as a function of wavelength.

Let's make a thought exercise. Suppose we have a perfect instrument, which offers the same sensitivity at each wavelength. Suppose also we have a perfect light source, emitting exactly the same energy at each wavelength. In this case, we would have a perfectly flat spectrum (fig. 9.3).

Reality is, of course, very different. Your instrument, taken globally (i.e. including the telescope and the acquisition camera) has a very wavelength-dependent sensitivity: this is the instrumental response. We talk about response, because we consider the instrument as a system into which we input a signal (the light) and which *responds* with an output (the recorded spectrum). Each element of the instrumental setup has its own response: telescope mirror, optical lenses (with the anti-reflection coating), diffraction grating

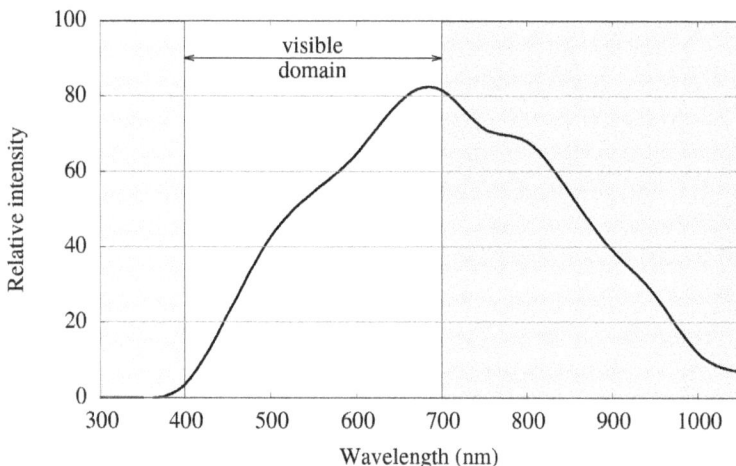

FIG. 9.4 – Typical response curve of a CCD.

(or prism), CCD cameras. To first order, the dominant contributions are from the grating and the CCD.

Generally, a CCD covers slightly more than the visible (especially in the near infrared), and its response curve is bell-shaped, like the one shown in 9.4.

Of course, CCDs with an extended sensitivity are preferable, with the condition that the sensitivity must also be high. Unsurprisingly, the best CCDs are also the most expensive. In any case, with amateur equipment, we never go below 350 nm or above 1,000 nm.

To determine the instrumental response of our instrument, it should be sufficient to shine onto it a "perfect" light (in the sense mentioned above: same intensity at all wavelengths), and measure what the instrument outputs. This is generally the procedure to characterize any instrument.

However, such a perfect light source does not exist either; the next best standard available has a Blackbody spectrum of the highest temperature possible. Now, a tungsten lamp has typically a color temperature below 3,000 degrees. When we trace the Planck curve of a light source at 2,800 degrees, for example (see section 2.3), we see that there is very little energy in the blue part of the spectrum (fig. 9.5).

This is a real difficulty for instrumental characterization!

The best solution available today is to look for a reference light source...in the sky! Stars are marvelous sources of light of high temperature, and if we choose a hot star, it offers a spectral distribution close enough to the Planck profile for our purpose, with just some absorption lines. For many stars, very precise spectra are available in "spectral libraries", and they can be used as reference spectra.

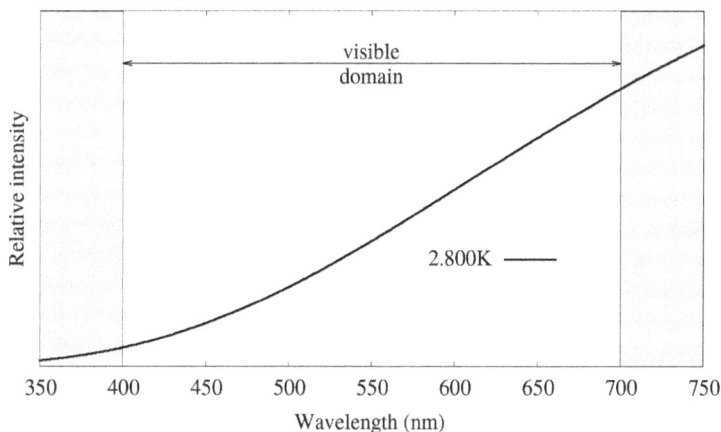

FIG. 9.5 – spectral distribution of a flat lamp (Planck profile at 2,800 degrees).

In practice, this means that for any given configuration of your imaging setup, you have to start each observing session by targeting a reference star, and compare the obtained spectrum with the reference spectrum of that particular star. By dividing the target spectra by the reference star spectra, you obtain the profile of the instrumental response.

The operation of smoothing is a bit more delicate, but essential to have high quality results.

To complicate things even more, when we observe a star, its spectrum is perturbed not only by the instrumental response, but also by atmospheric transmission. This is also important when trying to make precision measurements. Atmospheric transmission depends on many factors, such as air pollution (variable from site to site), humidity (variable within days, or even hours), and the thickness of the atmospheric layer crossed (variable depending on the latitude of the target). Think about a sunset : the color of the Sun at the horizon is very different from the Sun at the zenith (or the highest position in the sky).

The instrumental response is usually very stable in time (i.e. the optical elements of the telescope and spectoscope, and the detector, that is, the instrument). However, since atmospheric transmission varies with time, it is potentially different for every observation.

The result is that, in order to make a precise correction of the instrumental response, you need to observe the reference star more or less at the same time as your target (to avoid atmospheric variations), and at the same altitude above the horizon (to have the same thickness of the atmospheric layer crossed by the light) and as close as possible to your target star. In practice, all these conditions are almost impossible to meet, and the procedure needs to be refined depending on the specific case: it is common to make only one observation of a reference star per night. For "playful" measurements without high scientific ambitions, a very basic profile of the instrumental response is enough; it is enough, for example, to compare qualitatively two spectra.

Scaling...

In all the operations up to now, we have only made relative intensity measurements. The intensity of the spectrum at any point is given in ADUs, which has no direct relation with physical units of energy. The maximum level (in ADUs) in a spectrum can be very high if the camera has a large dynamical range (up to 16 bits), and if we stacked a large number of images. To avoid dealing with large numbers, it is wise to scale the spectrum, dividing each intensity by a constant, so that the lower part of the continuum has a value of about 1. This makes the arithmetic simpler, but doesn't otherwise alter the distribution of the spectrum with wavelength.

...or Normalization?

I have used the term *scaling* to describe the operation that allows you to have "ordinary" values for the intensity. You can also commonly find in the literature (and some software) the term *normalization* for this operation. Beware, this term has a different meaning depending on whom you are talking to. In the amateur community, it is commonly used to indicate the scaling described above, but among professionals it means a very different operation. Professionnals use normalization to adjust the continuum profile of a spectrum so that it only shows with the emission and absorption lines. The continuum is adjusted to equal 1 everywhere. It is not necessary to do this in general, particularly if you have absolutely calibrated your data. So it's best to avoid doing this. Therefore, it is a term to use with caution, to avoid confusion.

Exporting in a Standard Format

At the end of the data reduction process, the only thing left is to save the spectrum in a standard format. Depending on the purpose, the format can change, and it is useful to be able to export in different formats at the end of your data reduction:

- the FITS format is the most complete: it contains a header with all the information necessary for a scientific use of the spectrum (date, target, coordinates of the observing site,...), complying to the BeSS specifications. If you need to save your spectrum in one format only, use this one, because it is the only one that allows for the re-computation of all the others. The only inconvenience is that this requires tools that can handle it. If you want to share the fruit of your work (online, for example), it is unlikely that your audience will have all the tools necessary however if you want your data to be available for subsequent analysis by others, including professional astronomers, FITS is the format to use. FITS is also the format you will need to use to submit your spectra to one of the growing number of online spectroscopic libraries which will archive amateur spectra;

– the PNG format is complementary to the former: it is an image that contains the spectrum in a very common (and simple) format – any computer can read this format nowadays (it is one of the most common for images). On the other hand, it does not contain the richness of information of the FITS header, nor each precise value. It is the ideal format for online sharing;

– the DAT format is very rudimentary: it is a text file containing the spectrum in the form of two columns of numbers – the wavelength and the intensity. This format can be imported into computational software (Matlab, Scilab, Excel, etc.). It is the entry door to process the signal with powerful tools!

9.8 Catalogues of Reference Stars

In order to compensate for the instrumental response, we need to observe a reference star and compare its reference spectrum with our observation. But where can we find the reference spectrum, or in other words, how do we know if a spectrum is sufficiently trustworthy to be used as a reference?

It is a complex issue, that I cannot solve here. But I can cite a few catalogues available, either on the internet, or in the data reduction software (specifically in the ISIS software).

The reference catalogs are generally of two types: those based on real observations, either ground-based or from space, and made with a special care to the photometry (intensity measurement), and those based on simulations, to generate theoretical stellar spectra. The former have the merit of being exact, but the latter can be applied to any star.

When choosing your reference star, you only need to consider its spectral type, since stars of the same type have very similar spectra.

In addition to the source of the reference spectrum, you also need to check that it is compatible with the resolution you are working at, and your spectral domain.

Here are some specific comments on the most widely used catalogs:

– Pickles: a catalog including 131 stellar types, complied from different sources, and published by A.J. Pickles in 1998. These are not synthetic spectra, and the observations come from different instruments. They are low resolution spectra with a very large spectral domain (from 1,100 to 10,000 Å);

- Miles Library: it is a catalog of real observations, made with a 2.5 m telescope at a resolution of 2.5 Å. The spectral range goes from 3,500 to 7,500 Å;

- UVES: it is an instrument on the VLT (Echelle spectroscope). It gives very high-resolution spectra, but of limited spectral range. In ISIS only a few UVES spectra are available, around the H_α line;

- Élodie: it is another instrument, installed on the T2m of the Observatoire de Haute-Provence (OHP). The domain is limited to 6,200–6,800 Å, at very high resolution (0.05 Å).

I wish to mention here the work of François-Mathieu Teyssier (amateur astronomer very active in the community[29]), who created a table of reference stars in each portion of the sky. In compiling this table, François-Mathieu has taken care to consider only stars with no interstellar reddening (i.e. stars whose spectral energy distribution is not altered by the presence of dust clouds between Earth and the star). It is a practical tool very useful to prepare observations.

The topic of reference profiles is complex, since no star is "perfect", and the same is true for any instrument used to observe them. You can do the exercise of comparing different reference spectra from the catalogs mentioned above: there are visible differences. A good approach to avoid this is to always use the same reference catalog.

9.9 An observation is a Set of Images

The raw images of the object you observe are of course the heart of your observation. They contain the essential information that you want to measure. But these images suffer from a number of important errors which you must correct. To correct these errors and calibrate your measurements, you need several series of complementary images (often called *reference images*): calibration, bias, flat, and dark. And you also need a profile for the instrumental response (which is not an image in the proper sense of the word).

For this reason, I recommend that you consider observations as a coherent set of images, and only if you have all these images can you maximize the data in your raw image, thanks to the process of data reduction.

Let me specify an important point: each observation is a group of images, but this does not mean that the reference images need to be taken again for each new observation. Most of the reference images can be reused in a large number of observations. For example, we can utilize a flat image for the whole

---

[29] François-Mathieu Teyssier's website : http://www.astronomie-amateur.fr/

observing night, and we can also use the same dark image for several observing sessions – since these depend only on the camera.

Since the errors are intrinsically due to the instrument (as an ensemble), it is easy to see that the set of reference images, and the process of data reduction depend more on your configuration than on the observation you want to make. Consequently, it is advisable to define a standard observing protocol – including the reference images and the data reduction – that you adhere to, no matter what kind of observations you want to do.

Such a protocol is not very common among amateur observers, while it is obligatory among professionals. I assure you that it guarantees simplicity and efficiency: simplicity because you follow it systematically (no need to think during the observations), and efficiency because it can easily be automated (allowing for a complete data reduction in a few clicks).

Chapter 10

First Spectroscopic Observation: The Sun

You now have a functioning and properly set up spectroscope. Furthermore, you have the tools to reduce the data of a complete observation. As a target for this first full observation, use the Sun, which can be done even without a telescope.

During the day, all non-artificial light comes from the Sun. Even indirectly, even when passing through clouds, daylight displays strong signatures of the solar spectrum. Moreover, such light is abundant and you can register spectra with a short exposure time.

All the procedures I describe here can be done with any spectroscope; the figures in this chapter have been obtained with an Alpy 600 and an Atik 314L+ CCD camera.

So, comfortably sit in a room illuminated by the Sun (the light *does not* need to be direct). Turn on your spectroscope camera, the acquisition software, and activate the cooling (it is not strictly necessary, since the exposure time will be short, but it is a good habit to acquire).

Turn the spectroscope slit toward the daylight, possibly with a diffusing element in front of it (e.g. a white sheet of paper). Its effect will be to produce uniform illumination across the instrument's aperture. This prevents local effects on the spectrum (the spectroscope is designed to get input from a telescope, i.e. a uniform beam).

Take some acquisition images, to verify that your instrument setup is alright (clear spectrum, horizontally aligned image). Chose the optimal exposure time to yield of about 80 % of your camera capacity. The "saturation"'ADU level of an Atik 314L+ camera is 16 bits, i.e. the maximum possible level is $2^{16} = 65,636$ ADU. Therefore, you need to choose the exposure time to have a maximum level in the spectrum of about 52,000 ADU.

Once you have found the optimal exposure time, take a series of solar spectra – say 15 images.

10.1 Reference Images

As we saw in the previous chapter, observations are not finished with the raw images of the observed object (the Sun, in our case), but need to be completed with several reference images. Take these images right after the raw images, so they correspond to the same setup and conditions of your target images.

Take seven (or more) bias images. To do this, close the spectroscope entry (making sure there are no light leaks) and take exposures with zero exposure time.

Then, take seven (or more) darks. Again, close the spectroscope entry, but now use the same exposure time used for the solar spectra (or longer).

Take seven (or more) flat spectra. This is slightly more involved: you cannot use the Sun, but instead you need a source with a continuous spectrum, as hot as possible. The simplest thing is to use a halogen lamp – but above all, avoid energy saving lamps which have spectra full of emission lines. The difficulty, at this stage, is to avoid any contamination from sunlight while still letting the light of the artificial source into the spectroscope. If possible, close the room shutters or curtains... and in the worst case scenario, wait for night! Make sure you chose an exposure time that gives roughly 80 % of the camera maximum ADU level.

Now take a spectrum (or more) of a calibration source. If you don't have a calibration lamp available, you can cheat...and use the solar spectrum itself (the images you just made a few minutes ago). What we need out of a calibration spectrum, is to identify a few lines at known wavelengths...and the solar spectrum is full of those lines! Of course, calibrating a spectrum on itself is a scientific heresy[30], but for the purpose of familiarizing yourself with the technique, this does not matter.

For completeness, you need a reference source to calculate the instrumental response. In the previous chapter I said you should use a reference star, which is not easy to find during the day. Again, we are going to work around the obstacle – since we can't jump over it – and use once again the solar spectrum itself. The solar spectrum is very well known and we can compare the spectrum obtained a few minutes ago with a "reference spectrum" from your favorite observatory.

Once you have gathered all these series of images, then you can unplug your instrument and start the data reduction.

10.2 Data Reduction

You are now going to use these images within your data reduction software package. In the following, I used the ISIS software[31], developed by Christian

---

[30] In fact, this is not totally a heresy. The Moore solar atlas (NBS) or the Utrecht atlas are standard – just avoid the strongest lines (Balmer, CaII, etc.)

[31] ISIS : http://www.astrosurf.com/buil/isis/isis.htm

Buil, but the same procedure can be applied with different software tools. I suggest you do the data reduction step by step at the beginning, but later you can automate everything.

In the last chapter (see section 9.7), I have described the sequence of operations for data reduction: I will exactly follow that sequence.

It is not my purpose to give here detailed documentation for the ISIS software (there are very good tutorials addressing this on Christian Buil's website). Nevertheless, here are some general tips:

- make sure you select the right instrument (Alpy 600) in the configuration tab;

- verify that your image names end by -XX, where XX is their ordering number: this is the best way of working with ISIS;

- verify that all images are in the folder specified in ISIS (configuration tab).

The first step is the pre-processing, i.e. correcting for bias, dark, and flat. For this, use ISIS to create the "master images" from the series of bias, dark, and flat images. Let's start with the bias frames (the easiest). You only need to take the median of the series of biases you took – we saw in the exercises on the camera (section 7.4) that this reduces the read-out noise of the camera. You can verify this again by measuring the noise in a region of a raw image, and in the same region of the median image (fig. 10.1 and 10.2).

In ISIS, go to the "master image" tab, and fill the fields in the "Bias" region (note that in ISIS bias images are referred to as offset images). The generic name corresponds to the root of the image series. For example, if your image series is called *bias-1.fit, bias-2.fit*, etc., the root is *bias*. Assign a new name for the output image – for example *Bias*. Then, press GO: ISIS evaluates the median image, and you can open it from the Image tab.

Proceed analogously for the darks. Now you also need to indicate the name of the master bias image you just created. To create the master dark, ISIS takes the median of all darks, and then subtracts the master bias – which represents the systematic shift in ADU caused by the camera. The result is that the average level of the camera should now be very close to zero, since what remains is only caused by the thermal noise.

At this stage, ISIS has an option for the detection and correction of hot pixels, to prepare their further correction. A hot pixel is a defective pixel that is more sensitive than the others. Normally, pixels should all have the same level in a dark image. In the tab "Make a Cosmetic file", chose an arbitrary threshold – for example 50 ADU – and start the computation. ISIS indicates the number of pixels it found with a level higher than the threshold: typically, a few hundred hot pixels is considered acceptable. If you are far from this value, change the threshold and start over.

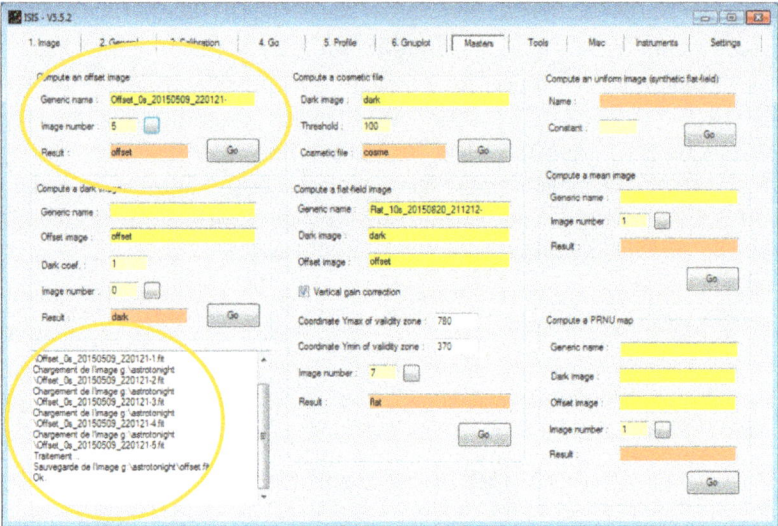

FIG. 10.1 – 2D Image: creation of the median of a series of bias images.

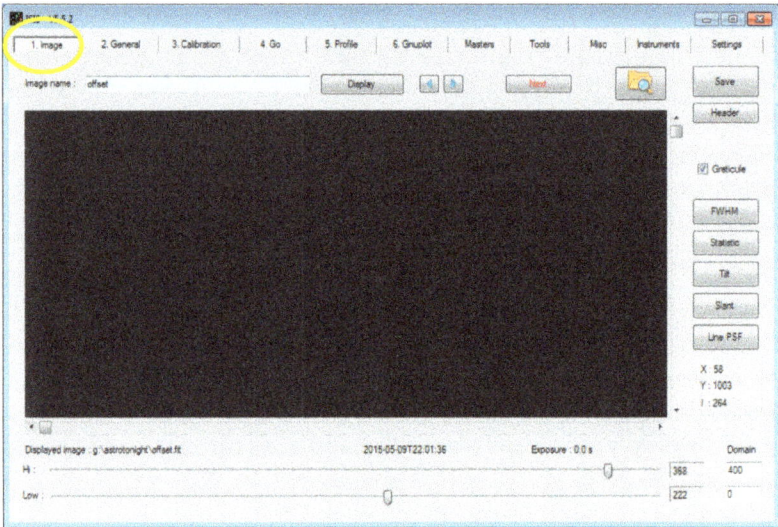

FIG. 10.2 – 2D Image: visualization of the bias median.

A better procedure is to measure the standard deviation of the ADU values in the master dark, and choose three times this value as a threshold.

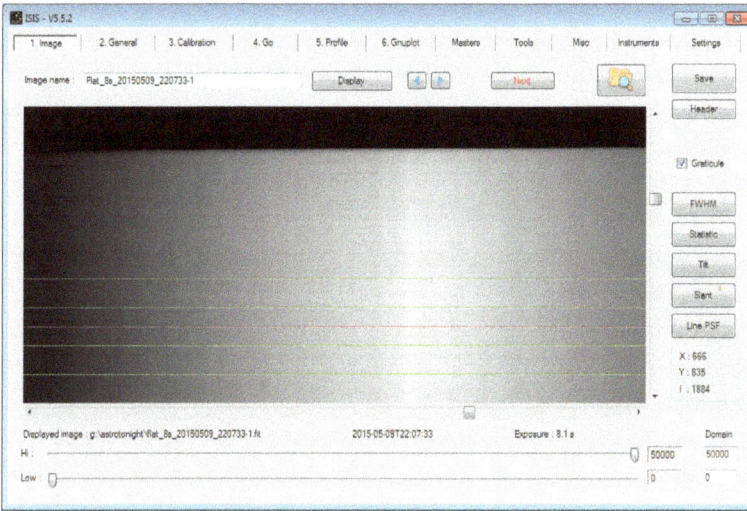

FIG. 10.3 – Flat image with slit gradient.

The result of this calculation is a file called cosme.lst, which you can open with your favorite text editor. You will find a list of values, each line corresponds to the coordinates of a pixel labeled as hot. You can check a few entries of this list on your image to confirm that they are brighter than their neighbors.

The calculation of the master flat is the next step. Ideally, the flat images should be perfectly uniform, but we cannot produce light with a perfectly flat spectrum. We are thus left with a very wide profile made of the continuous spectrum of the source, the instrumental response, and the intrinsic sensitivity of each pixel. Moreover, it is probable that the slit was not exactly uniformly illuminated when you acquired the flat images. This produces variations of the intensity in the vertical direction (along the slit) – the so-called *slit gradient* (Fig. 10.3).

Essentially, we are interested in the intrinsic sensitivity of each pixel (all the rest will be compensated with the correction of the instrumental response). ISIS compensates for the slit gradient when calculating the master flat. For this, it re-assigns the same average level to each horizontal line of pixels. You need to indicate the region containing the spectrum (minimum and maximum Y). Fill in all the fields and start the calculation, then open the resulting image in the Image tab (fig. 10.4).

You are now done with the master images: you can move on to calibration of the raw images. ISIS does a global processing on the ensemble of images (it is a very efficient tool for this), so you cannot separately perform each step of the data reduction. Nevertheless, we can follow the logic of these operations – ISIS does exactly what was described in the last chapter.

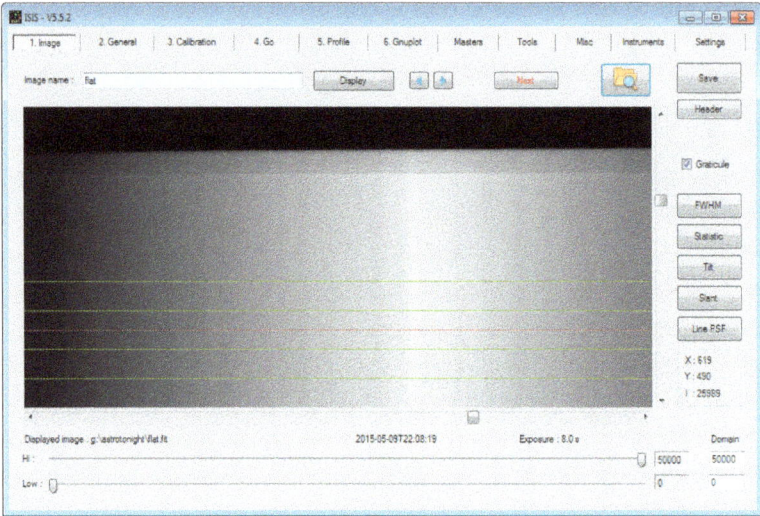

FIG. 10.4 – Resulting flat image.

ISIS allows you to follow the data reduction process by switching tabs from left to right. If you switch by clicking on "Next" (instead of clicking on the tab itself), ISIS automatically fills in. Go to the "Image" tab, and load the first raw image of the Sun.

Click on "Next" to go to the "General" tab (fig 10.5). You now have several fields to fill in, especially the name of your master images (bias, dark, flat, and possibly the cosmetic file).

Make sure to assign a name that reminds you of the target: Sun.

Check the "Sky Background not removed" option, since your solar spectra are for an extended source (there is no sky background on either sides of the solar spectrum).

Uncheck the option "Spectral calibration"; initially we are going to do the data reduction without wavelength calibration or correction of the instrumental response.

Click on "Next"; ISIS thus switches to the "Calibration" tab. Since we chose not to calibrate now, lets not spend time on this (we will come back to the calibration later). If you wish, you can set the "tilt angle" (rotation angle of the spectrum on the image) and the "smile" parameters (ISIS has tools to calculate them on the right hand side), but initially it is better to postpone these operations if your spectrum is relatively horizontal on the image. You can now click again on "Next". ISIS moves to the "GO" tab, and the only thing left to do is click on the GO button to start the data reduction. You can follow the calculations on the console, and after a few seconds, ISIS will show the messages displayed in figure 10.6.

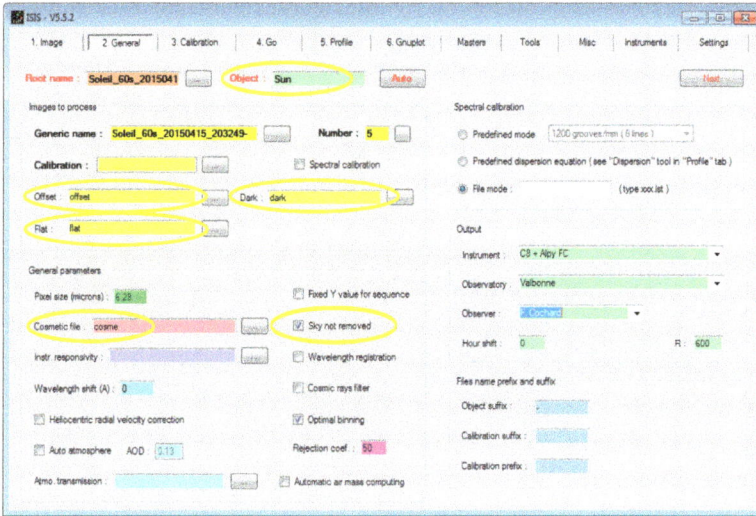

FIG. 10.5 – ISIS, general tab.

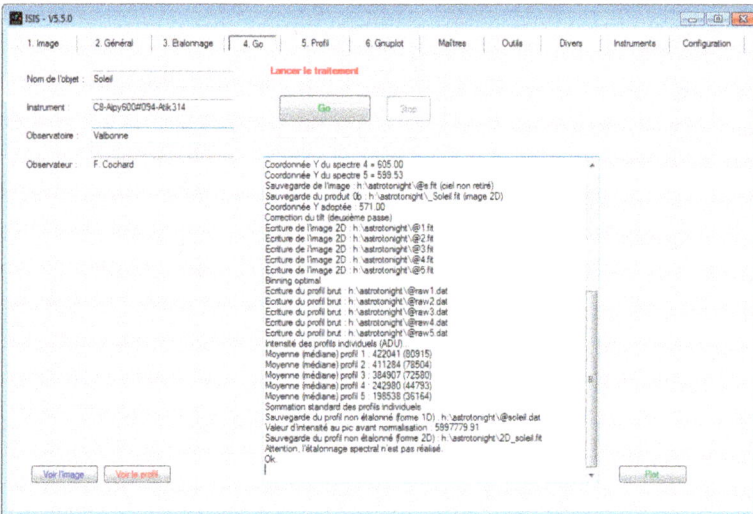

FIG. 10.6 – Results of the calculation.

At the bottom of the window, you can directly access the two main results of the calculation: one is the 2D pre-processed image, and the other is the extracted spectral distribution. Click on "Display image" first. Take the time to analyze this image, measuring its background level (close to zero), and noise (compare it to the raw images). Then, click on "Display profile" to look at the spectrum obtained by ISIS (fig. 10.7). It shows the intensity of the spectrum in each pixel. Since we haven't calibrated the wavelengths yet, the x-axis is in pixels.

You can immediately locate the main lines of this spectrum: the Sun is a star of intermediate temperature (or even rather cold for a star, since its surface temperature is of order of 6,000 K). Therefore, its spectrum shows a lot of absorption lines – more than 30,000 have been identified in total!

Let's take a step back. The three main operations of data reduction have been completed by ISIS in a few seconds: the pre-processing (including the correction of hot pixels), the geometric corrections (if you specified the tilt and smile parameters), and the optimized extraction of the spectrum. The calculation only takes a few seconds, but thousands of operations have been performed to deliver the spectral distribution you are looking at.

Nevertheless, two important steps remains before these data can be used: wavelength calibration, and the correction of instrumental response.

10.3 Wavelength Calibration

At the current stage, the x-axis of your spectral profile is in the pixel number of the CCD: this is not very useful physically. We now have to associate to each pixel a wavelength (in nm or Å).

Since the aim here is to make a first educational observation, we will not go too far in looking for a calibration source: the Sun itself is good enough. It offers a lot of easily recognized lines, and we are going to use this information.

In ISIS, when you look at the spectral profile, you have a tool to calculate the dispersion law (the button "Dispersion" on the "Profile" tab).

This tool provides several options to assign the wavelengths of known lines, and then measure in the spectrum the effective position of these lines (in pixel units). The lines can be in emission or in absorption (those in the Sun are all absorption). The tool also allows for the choice of the degree of the polynomial being used. Here, we can be happy with a fourth degree polynomial.

In the solar spectrum you can easily recognize, for example, lines of the following wavelengths (fig. 10.7 and 10.8):

wavelength	line
3,933 Å	K line of calcium
3,968 Å	H line of calcium
4,307 Å	G-band (molecular)
4,861 Å	Hβ (Balmer line)
5,167 Å	Magnesium triplet
5,270 Å	Iron line (Fe)
5,892 Å	Sodium doublet
6,563 Å	Hα (Balmer line)

Sun - raw profile

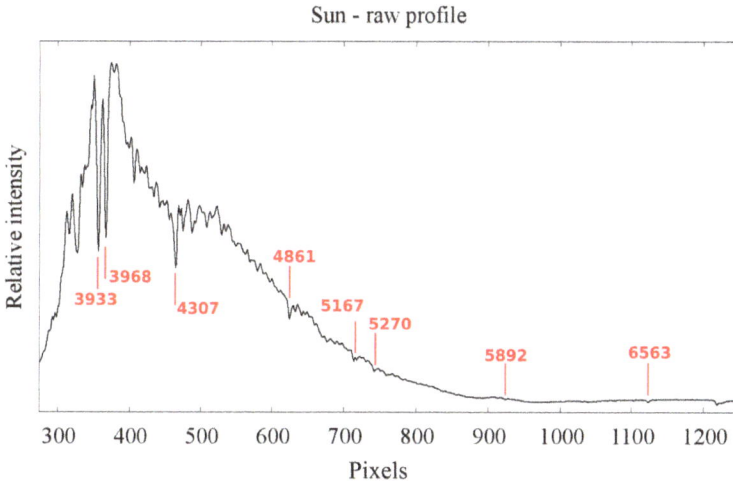

FIG. 10.7 – Spectral profile of the Sun, with some marked lines.

Manually choose these wavelengths. Then, for each line, double-click in your profile right before and right after the line: ISIS performs a very precise calculation of the line center in the interval you just defined.

When all values are set, click on the "Compute polynom" button. ISIS evaluates the polynomial of fourth degree fitting the data you provided, and ends the calculation evaluating the error (based on the measurement of more lines than the polynomial degree). This is the "RMS" value, corresponding to the typical difference between the real measurements and the calculated wavelength dispersion law. The RMS value gives the precision of the wavelength calibration. This precision must be consistent with the precision of the measurement of individual lines in your spectrum. If, for example, you evaluate that the position of a line can be measured with an accuracy of 0.1 nm, the RMS value given by ISIS must be, at the most, of the same order (possibly smaller). In the case of an Alpy 600 spectrum, the error is generally smaller than 0.1 nm.

If the error is greater, there can be several causes: either you have a problem in selecting a line, either you have a typo in the wavelength, or – the most common case – you selected the wrong line! It suffices to correct the erroneous value and re-launch the calculation. Note that ISIS also gives the difference between the measured line and the theoretical prediction for each line: if the error is large on a specific line, it is probably that one causing the greatest difference.

You can immediately apply this dispersion law to your spectrum (and from now on it will have the wavelengths on the horizontal x-axis), but you can also keep the spectrum as it is, and once again reduce your data: check the

"Spectral Calibration" box on the "General" tab and activate the "Go" tab again to produce a new and calibrated profile.

The dispersion law is characteristic of your specific configuration. One can consider that it will be *almost* unchanging. That is, as long as you don't modify your instrument (taking out the camera, re-setting the spectroscope, etc.), you can continue to use the same dispersion law. Nevertheless, I say "almost" because special attention is still necessary, at least in some particular cases. For example, if you work with a Lhires III in high resolution, a deformation of the support during telescope movement can alter the calibration by moving the spectrum by a fraction of a pixel. As it is often the case, everything depends on the precision you need for your measurement.

In most cases, the dispersion law itself is not modified (it only depends on the configuration of the instruments and the optics), but in some cases it can be shifted. In these rare cases. ISIS allows shifting the spectrum based on a specific line, without re-calculating the entire dispersion law (modifying the zero point).

Because of the risk of a change of the dispersion law during observations, I advise taking calibration images regularly during the night – and the more precision you need for the wavelength, the more images you should take.

ISIS offers several methods for dealing with wavelength calibration, each corresponding to a specific instrument or observation. For example, if you have a "calibration module" for the Alpy, you just need to indicate in ISIS the position of a single easily recognizable line, and the software recognizes the calibration spectrum on its own (specifically, it will be the spectrum of a mixture of neon and argon). Do not hesitate to explore these different options and chose your favorite based on the quality of the results and the ease of the procedure.

In our specific case of the Sun, we have slightly cheated for the spectral calibration, since we have used the spectrum of the Sun itself and took advantage of the fact it is a very well documented source in the literature. But the basic principle remains the same in all your observations: you take a calibration spectrum, identify a certain number of lines, use these to obtain the dispersion law and check the quality of the result (RMS error). This law is then re-used as long as the instrument has not changed.

10.4   Correction for the Instrumental Response

Once the spectrum is calibrated in wavelength, you still (in most cases) need to correct for the instrumental response. We saw in the last chapter (section 9.7) that for stars, the response depends not only on the instrument, but also on atmospheric transmission. Separating these two component is complex, and in practice we will not do it: it is best to use a known star (i.e. one for which we have a reliable spectrum in the literature), close to the target (say, less than 5 degrees of angular distance), and infer from this observation the instrumental response. By proceeding in this way, we consider the atmosphere as part of our instrument.

The standard procedure to obtain the curve of the instrumental response requires performing the entire sequence for reduction of the data of the reference star, and then dividing the extracted spectral profile by the reference profile:

$$\text{Instrumental response} = \left(\frac{\text{Obtained profile}}{\text{Reference profile}} \right)$$

Dividing one curve by another might seem strange but it is a simple operation: for each value of the x-axis (i.e. each pixel column) you divide the intensity in the extracted spectrum by the intensity in the reference spectrum. We can thus build point-by-point a new profile corresponding to the division of the obtained spectrum at each point by the reference spectrum – and generally we talk about division of one profile by the other.

Once the instrumental response is obtained, you can use it to correct all the subsequent spectra with the inverse operation:

$$\text{Corrected profile} = \frac{\text{Obtained profile}}{\text{Instrumental response}}$$

ISIS provides all the necessary tools to build the instrumental response. It also offers a large database of stellar spectra at different resolution: this saves a lot of time in finding a reliable reference spectrum.

We are going to use these tools with our solar spectrum.

Again, for simplicity, we are going to use the spectrum of the Sun itself as reference. The general distribution of the solar spectrum and the characteristic lines used above are very well known in the literature.

Of course, it is generally preferable to use hot stars (whose profile is smooth and with few lines only) for the reference spectrum, but we simplify things here, for practical purposes.

Initially, display the spectral profile of the Sun, calibrated in wavelength. You should see (with an Alpy 600) something like figure 10.8 (the general shape of the profile can vary a lot depending on the illumination conditions).

Now, use the tool "Database" in the "Profile" tab of ISIS (top right button) (fig. 10.9).

Many catalogs can be chosen, the one we want now is Pickles (top left). Chose from the list the profile of a G2V star, and load it in the profile window: you now have a profile like the one in figure 10.10.

The least we could say is that it is different from the profile you just obtained! Nevertheless, identify some of the main lines: you will see that they are present in both profiles. It is the general shape of the spectrum, which is deeply modified by the instrumental response, not the lines.

Save the profile from the database in your working directory.

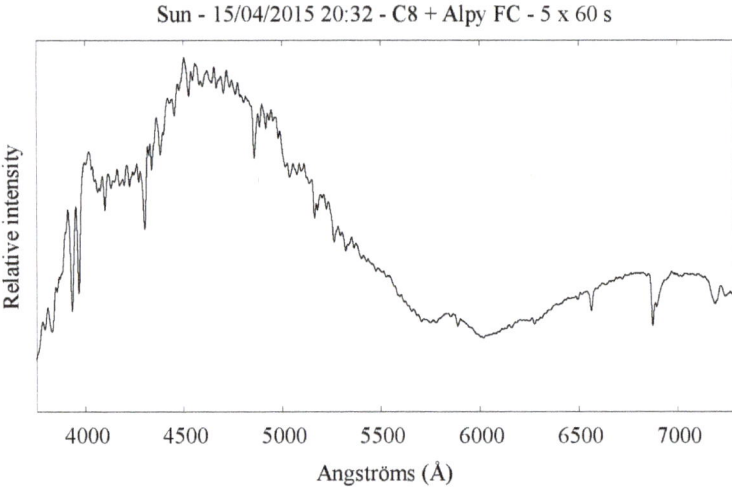

FIG. 10.8 – Spectral profile of the Sun.

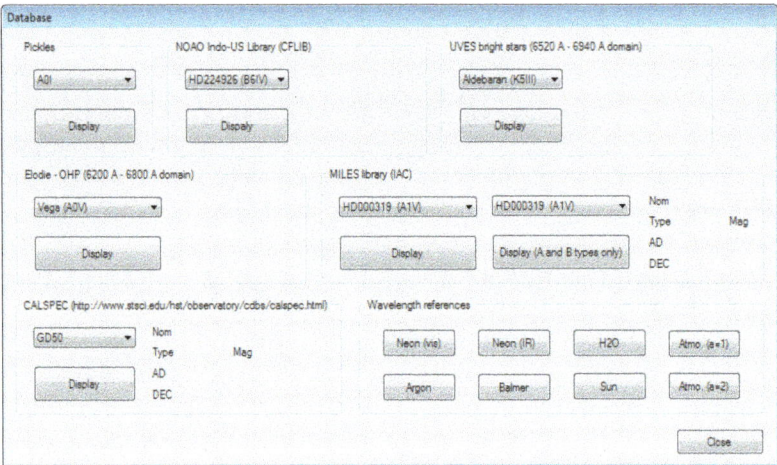

FIG. 10.9 – ISIS database.

ISIS offers a new tool at this stage: the "Response" tool. Select it (right column in the Profile tab), and indicate the name of the reference spectrum. The two spectra are shown in the window, and you can even carry out the division. You should then obtain an overall bell-shaped curve, but with a lot of fluctuations (green curve in fig. 10.11).

This "noise" comes from the very local irregularities in your spectrum. It is not, strictly speaking, part of the instrumental response – this changes only on a rather large scale (we say it is "low frequency").

Sun - raw profile

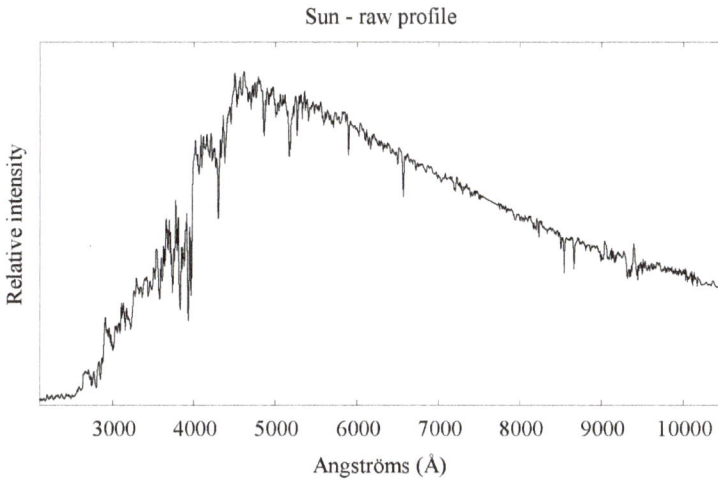

FIG. 10.10 – Reference spectrum for the Sun.

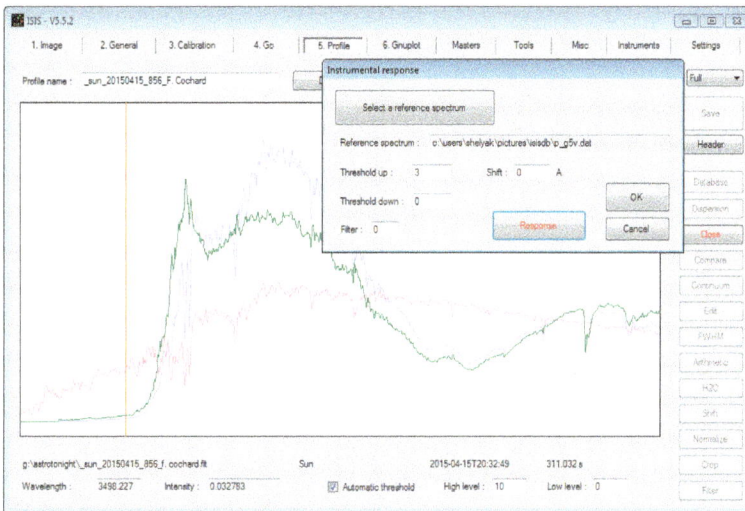

FIG. 10.11 – Tool for calculating the instrumental response.

Save this profile, then select the tool "Continuum". This tool allows you to smooth the curve, i.e. to eliminate the noise coming from our spectrum. You need to find the right settings to eliminate the noise ("high frequency") without altering significantly the general shape of the profile. In some places, there is a significant effect of deep absorption lines. These lines too do not

Instrumental response curve

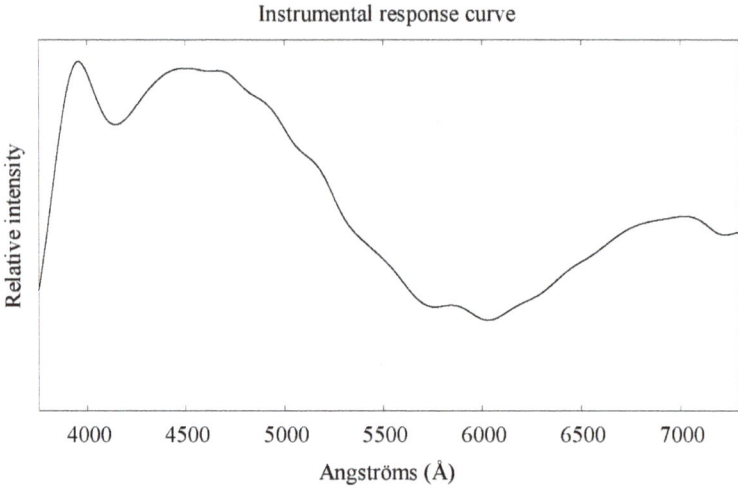

Fig. 10.12 – Calculated profile of the instrumental response.

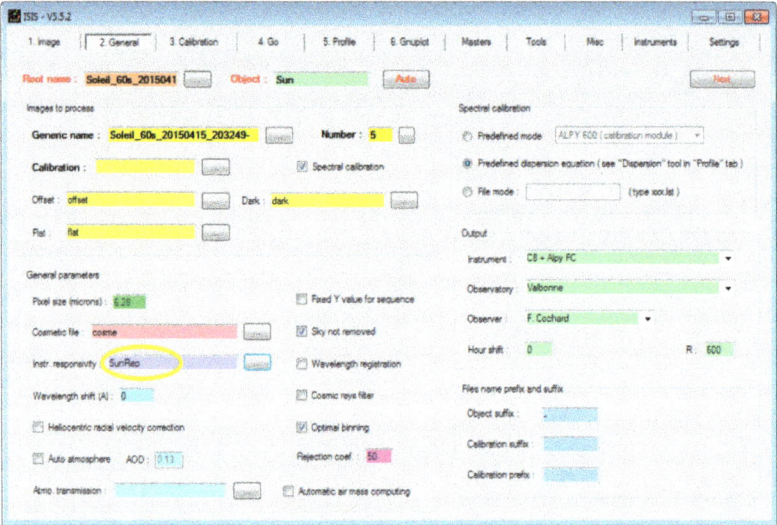

Fig. 10.13 – Instrumental response correction.

belong to the instrumental response, and it is better to eliminate them – the ISIS "Continuum" tool also provides this choice.

After removing the lines and smoothing the spectrum, you finally have your profile of instrumental response (fig. 10.12).

With this instrumental response, you can divide the solar profile (use, for example, the "Compare" tool of ISIS), but you can do better: go back to the General tab, and insert the name of your instrumental response profile in the "Instrumental Response" field (fig. 10.13).

Then re-launch the data reduction – this time, ISIS re-starts from the raw images, and follows the entire data reduction sequence, and in the end you obtain a completely usable spectrum: it is calibrated and the instrumental response has been applied. Now you can compare your result to the reference spectrum: the agreement is astonishing. You might say this is obvious, since we used the observed spectrum to correct itself. Sure, we cheated, but I assure you that when you perform the data reduction on a real star, you will have the same quality results.

The exercise we have done here with the Sun showed how to complete an entire observation on an easy target (i.e. which does not require a telescope). The main points illustrated by this exercise are:

- an observation is a coherent ensemble of images;

- data reduction is a complex operation, but it can be done in a few clicks if the reference images are adapted to the needs of the data reduction software;

- the data reduction process is specific for your instrument.

Chapter 11

Mastering the Telescope

Let's put aside for a moment the spectroscope, and focus on the other essential element of our installation: the telescope (which might as well be a spyglass).

If you already have experience with astronomical observations, and in particular if you already do imaging with a CCD (or a DSLR), most of this chapter will be familiar. Nevertheless, pay attention to some parts that are probably new to you – for example, how to place a faint star (invisible to the naked eye) at a specific location on the image, with pixel precision. In spectroscopy, the guiding field is often very small, and nothing looks more like a star than another star.

The first steps outlined here, up to the preliminary alignment, must be done during the day. It is useless to wait for night – darkness must be dedicated to observations, not to instrumental setup.

11.1 Mastering the Mount

It might seem trivial to say that one needs to master the mount, but experience shows that novice observers tend to jump over this step. One might assume that, having a "GoTo" mount, the telescope will find its way on its own, but unfortunately this is not true! A telescope mount, with its remote control, and PC connection is a complex piece of equipment with many functions. During a spectroscopic observation, you will need most of these functions. Be it the pointing, motion speed, alignment, mount reverse, PC control, etc., take the time to master each function. You need to be able to point to any object without consulting the manual every time! You need to be able to do this with the control pad and also remotely through the PC. In the following steps, we are going, little by little, to use all these functions.

I assume that you have installed your mount (with the telescope) on stable ground, with the right ascension (RA) axis roughly oriented north (worst case scenario: use a compass) (fig. 11.1).

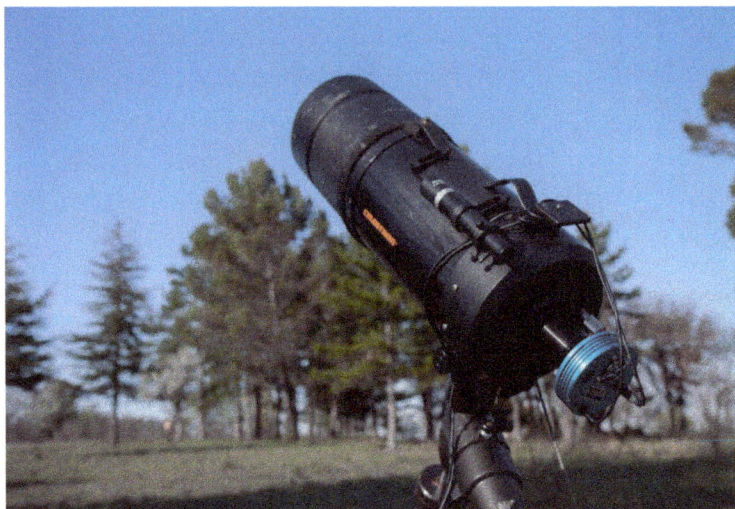

FIG. 11.1 – Telescope installation on the ground, looking north.

Install a CCD Camera

If your mount is completely new to you, I strongly recommend doing the first steps (alignment of the finder, alignment of the telescope, pointing) visually, through an eyepiece. But as soon as you are able to do these operations visually, you will have to start over with a CCD camera installed at the focus of the telescope. In fact, the field covered by the CCD sensor is much smaller than that of the eyepiece and the expected precision is therefore greater.

The problem of pointing and guiding the telescope will be dealt with later, so I suggest using your guiding camera from now on; of course, its field of view is probably smaller than that of your acquisition camera, but the other tools (acquisition software, PC connection) will rely on it to take spectra later.

So, install your CCD camera on the telescope (fig. 11.2). Make sure it is rigidly fixed to the telescope. When the telescope moves, the relative position of the camera must not vary. It could move, for instance, because of its own weight, or because of the cables. A motion of a few tenths of a millimeter translates into a motion of the image of several tens of pixels. It is a good idea to attach the camera cables to the body of the telescope so their weight doesn't drag on the camera.

Balance of the Mount

Make sure the mount is well balanced (in both axes), that is, even when you loosen the brakes, it does not swing (fig. 11.3). This balance is important to prevent the drive motor from carrying all the weight of a possible

FIG. 11.2 – Camera mounted on the telescope.

FIG. 11.3 – Balancing the mount.

imbalance: the motor is not designed for this. The balance is achieved differently for each mount type, but generally speaking, you have to move some weights or the telescope itself on a sliding track.

To test the balance, perform the following operation on each axis:

- Release the brake (be careful to avoid anything falling);

- Make the mount move (manually) in both directions. The motion should have roughly the same freedom in both directions, and smooth, without any blocking point;

- If this doesn't happen, move the weights or the telescope again and proceed by successive iterations.

- Re-apply the brake.

Put yourself in the motor's shoes for a moment, and ask yourself if the effort required to move the mount is reasonable.

Rough Set Up

When the telescope is ready, plug in the equipment and start the acquisition software for your camera. Point the telescope to the farthest target possible (on the horizon), and start continuous acquisition with a very short exposure time (since there is still daylight, there is a lot of light. Set the exposure time so that the image is not saturated – if necessary cover part of the telescope aperture).

Then, roughly focus, so that you can easily distinguish the field observed by the telescope. A rough set up is sufficient, since you will need to refocus once darkness falls to observe stars.

Alignment of the Finder

The field of view of the guiding camera is small, and it is not easy to put a star in this field. To bypass this difficulty, your telescope has a finder, a small telescope with a moderate magnification installed in parallel to the telescope (fig. 11.4). Since the field of view in the finder is much larger than for the telescope, it is easy to "catch" the target.

It is possible that your finder isn't a small telescope, but a slightly different device: a laser pointer, or a *Telrad*. The principle is the same for the alignment procedure.

It is unlikely that the finder is perfectly aligned with the telescope when it is new, or it has traveled. If this alignment is imprecise, you can put the target star in the finder field, but it will not appear in the image taken by your camera. Take the time to verify this alignment and correct it if necessary. Do this before twilight. Find a typical object on the horizon (tree, mountain, electric pole, ...) and look for it with the telescope – the surroundings of your target will be helpful. Then, when your target is well centered on your image, align the finder so that your target is perfectly in the center of the finder's crosswire. Once this is done, during the rest of the night you will just have to put the target star at the center of the finder's crosswire to see it in the center of your image.

Once the set up is satisfactory, wait for night and successively point at several bright stars with the finder, using the telescope remote control: they must be at the image center.

Alignment of the Mount

The mount alignment consists in aligning the mount's polar rotation axis to the Earth's rotation axis. This allows the telescope to precisely follow the sky's rotation. The exposure times in spectroscopy are often long (from few

FIG. 11.4 – Alignment of the finder.

minutes to hours), during which the star (whose image is just a few microm-eters at the telescope focus) must stay centered on the slit (which is also a few micrometers). This requires special care be taken to achieve good mount polar alignment. Of course, we will see later that the precise tracking of a star is done with the autoguiding, but this must be a complement to the mount alignment, and not a compensation for a bad polar alignment.

It is probable that your mount allows for a "software" alignment, i.e. that it requires you only to point to three stars and it then finds the direction of the Earth's rotation axis on its own. This method is useful for visual observations, but clearly insufficient for our aims. If the polar axis of the mount is not aligned to the Earth's rotation, the mount needs to use both motors with different speeds to follow a star on the sky. In our case, we need a real physical alignment of the mount's polar axis to the terrestrial rotation axis – so that later, tracking of the star happens on the right ascension axis only, and with constant speed.

Your mount might have a bubble level to check for horizontal alignment. I have often seen observers convinced that a good alignment can be achieved only with a perfect horizontal alignment. That's not true! The horizontal alignment is useful only to rapidly go back in position when you have to set up the telescope often (i.e. when the telescope is not always in the same location). Such a bubble level is also useful to avoid excessive inclination of the mount base – and prevent everything from overbalancing when the telescope is in certain position. But other than these (good) reasons, there is no need to align the mount base horizontally. Only aligning its main (RA) axis to the Earth's rotation axis is necessary.

The first step occurs during the installation of the mount on the ground: make sure the rotation axis for right ascension (RA) is pointing northward. The precision must be at most a few degrees (easily achieved "by feel").

There are several methods to align the mount – you should easily be able to find the documentation for each.

The easiest, if your mount has this option, is to use the polar scope. It is a small telescope integrated in the mount and aligned to its rotation axis. It is sufficient to point to the North star with this scope (which usually has a rotating crosswire). The precision is not astonishing, but it is generally sufficient for our purpose. This method has the advantage of requiring very little equipment, since it is visual.

The second method available for polar alignment is the King method. It is a smart – and old! – method, which consists of pointing to a star near to the North star, and looking at its displacement in the focal plane (i.e. on the image produced by the CCD). Depending on the direction and speed of this motion, you can define the corrections to the azimuth and height of the mount. This method, like the polar scope, requires being able to see the North star.

> To improve the precision of the alignment with my polar scope, a long time ago I did an alignment with the King method, and then I marked the position of the North star on my polar scope. By finding that precise position again at each new set up, I systematically recover a very satisfactory alignment.

There are cases when the polar scope cannot be used, especially when the North star cannot be seen from your observing site. In such cases, you need to use the Bigourdan (drift) method: it is the third method, and it is also old. Start by observing a star close to the meridian and observe the motion of its image in the focal plane. If the polar alignment is not perfect, the star moves vertically (along the declination axis). The correction required must be applied on the azimuthal axis of the mount. When the alignment is satisfactory, move on to observing a star at west (or east) and observe the motion of its image again. Now the correction must be applied to the height of the RA axis of the mount.

With a bit of experience, the alignment can be done in a few minutes (using the polar scope) or a few hours (for the King and Bigourdan methods, the longer you wait, the better the precision of the alignment). For this reason, the polar scope should be used if you often observe from different locations. If your installation is fixed, it is worth taking half a night to do a very precise alignment.

Precision of the Alignment

As for all iterative methods, one must ask the questions: "how far should I go? What precision do I need to achieve?" In the case of the telescope alignment, these questions are even more tricky, since later we will have the autoguiding on, which will compensate for a good deal of the errors in the polar alignment. Thus, it is tempting to do a quick and dirty alignment and let the autoguiding do all the work. I do not like this approach: the less you ask of the autoguiding, the better it works, and the easier it is to set it up.

In practice, I consider the polar alignment to be satisfactory when a star stays in the same position – say within a few pixels – for at least 5 minutes...without autoguiding! With a little practice, this is easy to achieve.

Conversely, if you are a beginner, do not spend the whole night tweaking the alignment: only once you've taken your first spectra you can have a feeling for the precision of the alignment you need.

During all your observations, keep in mind that if a star moves (too much) on your guiding image, the cause is probably a problem in the polar alignment of the telescope. Nevertheless, beware: if the alignment was perfect for several hours, and suddenly a star starts moving on the image, check that you don't have a cable stuck which slows the mount (true story!). If this is the case, you probably want to shut down the motors as soon as possible.

Initialization of the Mount

Once the polar alignment is good, you still need to tell the mount where it is. Most amateur mounts do not have an absolute reference, and once plugged in, you need to give a reference position. Since the mount moves along two axes (right ascension and declination), it is sufficient to point to a star of known coordinates, put it in the center of the image, and tell the mount the coordinates it is pointing at (i.e. the coordinates of the star).

From now on, the mount should be able to point to any object of given coordinates. It should...the conditional tense here is important: we will see that, often, things are not as simple.

If your mount has a "parking" function, you should definitely use it at the end of your observations: it is a simple way of putting the mount in a known position, so that you do not need to give it a new reference next time you plug it in.

Beware: your mount probably has several functioning modes – in particular, on top of the "polar alignment mode" we just went through, it can certainly also function in "non-aligned mode". In this case, use the "software" alignment, already mentioned above: it requires pointing to three stars. I have often seen observers doing a physical alignment of the mount and then use three reference stars to align their mount, assuming that it must be more precise than just one. This is wrong: the two functioning modes are different and incompatible. If the mount is well aligned to the Earth's rotation axis, one star is enough to make it understand where it is pointing, while if the mount is not well aligned, it needs three stars to figure out where it is looking on the sky. Let me remind you that for spectroscopic observations the former mode (physical alignment of the mount) is to be preferred.

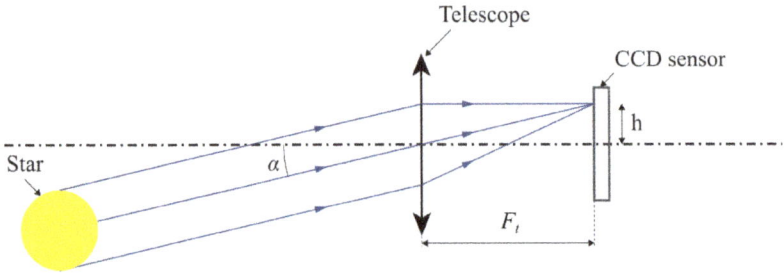

FIG. 11.5 – Calculation of the position of the image of a star.

11.2 Calculation of the Image Field

In many situations, you will need to know the size of the field of your guiding image. For example, whenever you look at a sky map and you want to compare it to your guiding image: how do you know how much to zoom in the map? It is already hard to recognize a field of stars by comparison with a map. If, on top of this, you don't know the scale of the image, it becomes a gambling game.

Therefore, it is useful to know how to measure and/or calculate the field in your image. Several methods are available.

Theoretical Calculation

The first method is a theoretical calculation based on the focal length of the instrument and on the size of the detector. We can consider the telescope as a single simple lens of focal length F_t, which projects the image of a star (assumed to be at infinite distance) on the CCD surface as in figure 11.5.

The image of a star at an angle α from the telescope axis will therefore be at distance h from the center of the detector. The relation between α, h, and F_t is the following:

$$h = F_t \times \tan(\alpha)$$

We can use this formula for stars at the edges of the detector and infer the maximum angle observable on the sky (calculated from the optical axis of the telescope), which is the field in the image (fig. 11.6).

$$\alpha_{max} = \arctan\left(\frac{H}{F_t}\right)$$

Where H is the total height of the detector.

Beware: here, I am using the height of the detector, but CCDs are typically not square. One can repeat the calculation for its width, or even its diagonal.

Measurement using a Known Stellar Field

A second method to define the field in your image is to point the telescope toward a known field and find two stars (A and B) of known coordinates (for example, using a stellar catalogue) in the image. By measuring the position in the image (X and Y) of each of these stars, you can use the Pythagorian theorem to infer the field per pixel. Thus, knowing how many pixels are in your CCD, you can infer the total field seen by the whole detector.

Measure by Moving the Telescope

A third method consists in using the telescope to do a similar measurement. Find two stars in your image (possibly far from each other), and measure their distance in pixels (using the Pythagorian theorem, exactly as in the previous case). Put the first star exactly in the center of your image (using the remote control of the mount), and take note of the position of the mount (coordinates in Right Ascension – RA – and Declination – Dec). Then put the second star in exactly the same position in the image and, once again, take note of the new coordinates the mount is pointing at. You have now all the data to calculate the angle between the two positions and infer the field per pixel, as in the previous method (but this time, without having to know anything about the two stars used).

11.3 Understand the Telescope Motion

Let me state the obvious once again: it is useful to take a few moments to understand how the motion of the telescope (along the right ascension

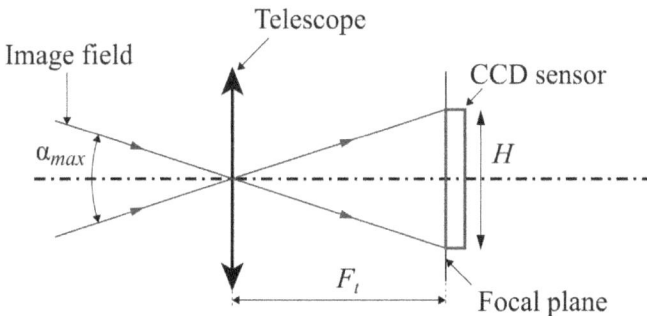

FIG. 11.6 – Calculation of the field of the CCD camera.

and declination axes) translates in the image. Intuitively, if the telescope is properly set, I would expect that stars move horizontally in the image when I move the telescope along the RA, and vertically when I move it along the DEC. In reality, all depends on the actual orientation of your CCD. If the camera is rotated by 90 degrees, you see the exact opposite – it is not necessarily a problem (you might have a good reason to rotate the camera), but you have to consider this when moving the telescope (for example, in this case, if you want to move the star vertically, you have to move the telescope with the RA drive). It is also possible that the camera has an arbitrary angle with respect to the mount: I strongly advice against it! In such cases, a motion in RA (for example) translates in a diagonal shift of the star...very hard to master, and it will have many drawbacks during the observation.

In some cases, you might deliberately choose such a configuration – for example to align the slit to the tail of a comet. This immediately complicates the observation: do not do it until you completely master your instrumentation.

Keep in mind that when you move the telescope (through the remote control) along the RA axis, and then along the DEC, stars must move in the image parallel to the edges.

Let me also highlight a common trap. Your mount certainly allows for different kinds of movement: along the X and Y axis of the mount (which is what we care about), but also, for example, along the geographical directions of the observing location (left/right and up/down) or along the sky coordinates RA and DEC (which differ from the mount X and Y if it is not aligned to the Earth's rotation axis). Make sure you work with the X and Y motion: it is the only mode which requires only one drive (RA) of the mount and relates to the image in your CCD camera.

Sky Maps

You will certainly need a sky map. Actually, nothing looks more like a star than another star, and to point exactly to the the star you target, you need to compare the observed field with the field on a map.

There are many software sky maps: Skychart[32] (fig. 11.7), C2A[33], TheSkyX[34], Guide[35], MegaStar[36], etc.

Let me also mention Stellarium, which is not as "technical", but it is very intuitive and aesthetic. It might be a good starting point.

These software packages use many stellar catalogs, and are generally capable of identifying any target. And in case an object is unknown to the software, it can query an online database (see below: CDS and Simbad).

---

[32] Skychart : http://www.ap-i.net/skychart/fr/start

[33] C2A : http://www.astrosurf.com/c2a/

[34] TheSkyX: http://www.bisque.com/sc/pages/TheSkyX-Professional-Edition.aspx

[35] Guide : http://www.projectpluto.com/

[36] MegaStar : http://www.willbell.com/software/megastar/index.htm

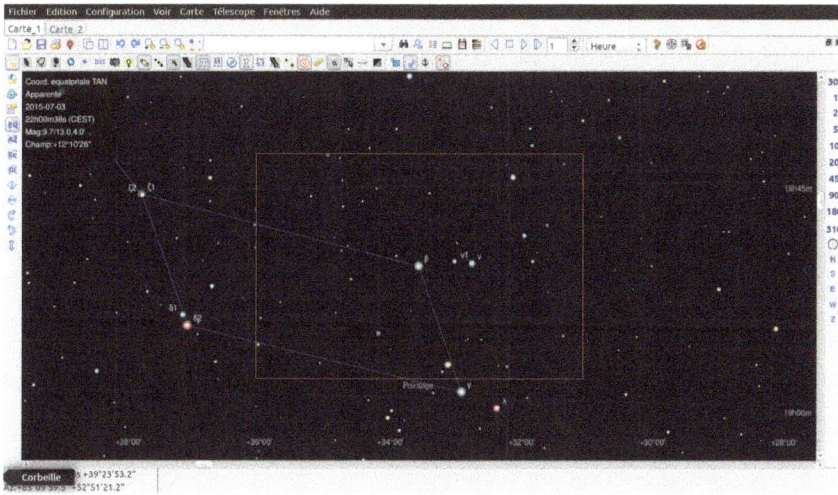

FIG. 11.7 – Screen shot of the software Carte du Ciel showing the field of the pointing image around βLyr. The red rectangle corresponds to the field of my CCD camera.

Take the time to chose one of these tools and familiarize yourself with it. Take the time to identify the observed field (CCD image) with the corresponding map field. It is a tricky exercise the first time, since you don't necessarily know the size of the observed field, or the orientation of the map with respect to the image (there might even be an inversion of the image). With a bit of practice, though, this exercise only takes a few seconds.

An important element: if your map software is well designed, it should be capable of showing the field either as a function of the local coordinates (the sky as you see it with the naked eye), or as a function of the equatorial coordinates (i.e. the axes along which the telescope can move). Therefore, it is possible to set things so that you see the image of the CCD and the map in the same coordinates: this greatly simplifies the task of recognizing the stars in the field.

11.4   Pointing to a Star

This will sound trivial...you need to know how to point to a specific star. This means aligning the telescope to that star so that it is in the center of the observed field (and later, in the slit of the spectroscope). You might think that with current telescope mounts this is easy, but this is not necessarily true. In the process of learning how to do spectroscopic observations, this is often a delicate step, for three reasons:

- the guiding field is typically small. One aims for a high magnification, to precisely position the image of the star onto the slit, but this comes at the cost of the width of the field;

- several elements can affect the precision of pointing: imprecise alignment, deformation of the mount or of the telescope, mechanical drag on the drives, atmospheric refraction, etc;

- finally, you need to be able to locate the object in the sky (and it might not be visible to the naked eye), and position it with pixel precision on the guiding field.

The typical situation is this. You enter the coordinates of your target and press "GoTo", and nothing shows up on the image. You start wandering, slowly moving the telescope along the different axes of the mount. Sooner or later, you will end up finding a star, but how can you know it is the right one? Don't worry, most of the astronomers I know, including the most experienced ones, have observed the wrong target, and it will happen to you also some day. Nevertheless, it is a pity to introduce a random element in the observation of an object whose coordinates are know to such extraordinary precision.

There are many ways to point to a star. The most basic is to spot the star with the naked eye, then point to it with the finder, and finally center it in the image (on the computer). If you are not familiar with this exercise, do it several times. This method is very efficient, but it is limited to the stars you can see with the naked eye, and possibly also by the finder.

Then, there is the "GoTo" function of your mount, using a star known to the remote control. Find a bright star and start the pointing. If your mount is properly set, the star you looked for will be close to the image center. But it is likely that it is not exactly in the center, because of all the reasons listed above. At this stage, there may be significant differences between an ordinary mount and a very good one – the difference is also in their prices. Therefore, you have to apply manual corrections (using the remote control) to put the target star precisely in the center. Such corrections are easy as long as the field of view is significantly larger than the positioning error and you can recognize this field. But when you do the same operation with the guiding field of the spectroscope, it is likely that the star will simply be outside the field of view.

The farther the target star is from the star you used to initialize the mount, the more important is the time elapsed since the initialization, and the greater the pointing error. For this reason, it is better not to switch from one side of the sky to the other too often (instead, organize your observations to only make a small movement each time). You can also re-initialize the mount (i.e. re-tell it what star it is pointing at – "syncing") each time you point a new star. This will reduce the error on the next pointing.

Let me make a side remark. To improve the precision of pointing, there is a very useful tool: the pointing model. It consists of pointing at many stars all over the sky, and saving each one of these pointings. The mount can then measure with great precision the pointing errors in all directions, which include all the issues described above. From this model, the mount can infer the effective direction to point at so that the target is actually in the center of the image.

You can, of course, use this technique to improve your pointing, but it is typically used only for permanent installations, since it takes a long time to do accurately.

Pointing without Fail

To avoid pointing at stars by trial and error, we need a reliable system to point without fail. Let me say it again: it is often a severe difficulty for beginners to point to the star of their choice.

Depending on your installation, there are several methods, which might even be combined together:

- I personally use a pointing camera which complements the guiding camera. It is a small camera mounted parallel to the telescope, with an objective of 70 to 100 mm, maximally open (if possible, with focal ratio F/2 or better). This camera observes the "big field" of the sky (a few degrees), and allows me to see maybe ten stars in a short exposure time (roughly one second). I compare the star field with maps, and once I recognize the field, thanks to the map that indicates precisely where the target object is, I can move the telescope until the star I want to observe is in the center of the guiding field. The focal length of this objective must comply with two constraints: the field must be large enough to cover the pointing errors (the better your mount, the smaller the field can be), but it must be small enough that the field of the guiding camera is large enough in the pointing image (a few ten pixels at least). Thus, when I put the star in that region, I am *sure* to see it in my guiding field. An important detail: at the beginning of the observation, I need to know where the center of the guiding field is in the pointing field. Since I took care to align my finder on the telescope before the observation, I only need to point to a bright star with the finder, verify that it is on the guiding field, and take note of its position in the pointing field. The relative position of the center of the guiding field on the pointing field will not change during the whole night.

- Most new mounts offer a "high precision pointing" mode. It is based on pointing first a bright object (i.e. visible in the finder) close to the target with the finder. When the star is in the center of the guiding field, you tell the mount which bright star it is pointing at. Then you use the "GoTo" function to point to the actual target star. Since the two stars are close to each other, the pointing issues of the mount are small, and your target will be in the guiding field without fail.

– There are fortunate and happy observers who own a mount with great precision, who carefully took care of the alignment (see above), and who also saved a pointing model. In this case, the mount can systematically put the target in the guiding field (and even in its center!). It is the ideal solution (and also the fastest way to point to an object), but such equipment is rare and expensive in the amateur community.

Each method has its own advantages and drawbacks. Let me draw your attention on one important point: the pointing camera, and the high quality mount allow you to point to an object without touching the telescope during the observation (and also without having to stand up from your workplace). Experience will show you that usually there are a lot of cables around the telescope and when you look into the finder, the risk of pulling a cable and "ruining" the alignment is high. In such conditions, it is is a necessity not to have to touch the instrument for the whole night.

Once again, I urge you to practice on your telescope, and point at fainter and fainter objects. For example, start with Vega, and then increase in magnitude until you reach the detection limit of the guiding field. Do not point randomly, always start by determining the coordinates of your target.

Finding the Coordinates of the Star

Once you know how to point without fail to a star with your telescope, there is still a common difficulty: how to find the coordinates of any object, especially the faintest. Usually, spectroscopic targets are not in the database of the mount.

If you are used to sky imaging, you probably rely a lot on the "GoTo" function to locate the target field. Typically, you are not looking for a specific star: what you want to image is a known and recognizable field (nebula, galaxy, planet,...). The problem is completely different when you want to point to a given star in a field of stars. For this, recognizing the field and identifying the target star are unavoidable steps.

The brightest stars are stored in your mount, and you just need to enter their name. The only complication might be the variety of names existing for the same object. There are many catalogs in astronomy, and each can give a different name to the same object. Bright stars have several tens of different names!

In our amateur activity, we prefer the Bayer catalog, which defines stars with a Greek letter, followed by the Latin name of the constellation. The first letter of the Greek alphabet (α) corresponds to the brightest star in the constellation, then β for the second brightest, and so on[37]. The star Vega (very common target for spectroscopist), for example, is called α Lyrae in this

---

[37] This not always true, but most of the time. In some cases, the order does not exactly match the brightness order.

catalog. In the files containing the spectra, we write the name with three small letters for the Greek letter (for example "alp" for alpha), and three letters for the constellation, with the first one capitalized. For Vega, this results in "alp Lyr".

For the faintest stars, not included in the Bayer catalog, we rely on the HD catalog (compiled by Henry Draper, hence the name of the catalog), which goes down to magnitude 9 or 10. The advantage of this catalog, is that it also gives the spectral type of each object. The name in this catalog is HD followed by a number. For example, Vega is also called HD172167. There is also the HR catalog (Bright Star Catalog) – Vega is HR7001.

CDS and Simbad

To find the correspondence between different catalogs and much more information on all stars, there is a fabulous tool on the Internet: the Simbad[38] tool from CDS. The CDS (Centre de Données astronomiques de Strasbourg[39]) is *THE* world reference for astronomical objects. It is the most complete database on the topic, and it is accessible online. Thus, in a few clicks we can access the Benedictine work of generations of researchers. With this tool, you can search for one or multiple objects by name or coordinates, or even do complicated queries to create lists of stars complying with multiple criteria. Start by choosing the option "by identifier". Type in, for example, "Vega" in the identifier field, and click on "Submit id". You then get a page giving the coordinates (in many reference systems), the spectral type, photometry (in many bands), the list of the different names for this object – 55 different ones for Vega! – and then a wealth of other information[40]. Of course, the brighter the star, the more it has been observed, and the more information is available.

Take the time to work with Simbad: it is a tool you will often go back to, almost each time you need to point to a new object. Once you are familiar with its interface, you will know how to immediately find any star, no matter what its name is.

11.5 Autoguiding

The autoguiding function puts the computer in charge of controlling the telescope and maintaining the star in the same position on the image – with pixel precision. As soon as you use long exposure times – and in spectroscopy the exposure time can exceed one hour –, you need to compensate for the inevitable deviations of the mount. Such deviations are caused by the same things complicating the pointing of the telescope: non-optimal alignment,

---

[38] http://simbad.u-strasbg.fr/simbad/
[39] http://cdsweb.u-strasbg.fr/index-fr.gml
[40] CDS also gives bibliographic links for the objects. There is also a complementary tool for bibliographic links : the ADS. See http://www.adsabs.harvard.edu/

mechanical deformations, atmospheric refraction, etc. The autoguiding is widely used for deep sky imaging, and the same rules apply for spectroscopy.

There is, however, one significant difference between guiding for deep sky imaging and for spectroscopy. In imaging, suboptimal guiding translates into non-spherical images of the stars – they are stretched. In spectroscopy, sub-optimal guiding translates into a loss of light into the instrument. There will be less light – and thus less information in the spectrum – but the intrinsic properties of the spectrum (resolution, calibration) do not vary.

Guiding the telescope can be manual – you, the observer, check the image and compensate if needed. I even recommend doing this for the first observations. This will give you a feeling for the compensation needed. But you will soon realize that staring at a star on an image is not the best use of your time. Moreover, experience shows that – except a few extreme situations where the human eye is better than the machine, the tracking is more regular and of better quality when automated.

The principle of autoguiding is simple. Aside from the main camera, which does the acquisition, the secondary camera, the so-called "guiding camera", checks the observing field at higher frequency – say, one image per second. The autoguiding software does a real-time analysis of these images, measures the exact position of the reference star and its distance to the target, and when this changes, it makes the mount compensate appropriately.

On paper, the principle is straightforward. In practice, there are many difficulties.

First, we have to connect a second camera to the computer. I have mentioned before the many IT problems frequently faced on the field, and this is the heart of what can be difficult. The connection of the camera itself might be very simple, but if you have chosen to use the same software to control all your equipment (e.g. AudeLa, Maxim DL, AstroArt), you need to make sure it can deal with two cameras at the same time, without driver issues. Asking a computer to deal with a stream of images is a very common thing nowadays. Asking it to deal with two streams of images (on top of which we have the real time analysis of one of the streams) is much more rare. The power of modern computers is more than enough to do this, but problems in the connection and recognition of the two cameras are common.

The second difficulty is to connect the telescope mount to the computer. There are many ways to do this connection, depending mostly on your mount. A standard "autoguiding port" has been around for a long time, it was developed by SBIG with the ST-4 camera. This port is a connector accepting commands similar to those of the remote control. In general, a small box between the computer and the autoguiding port contains switches allowing the computer to control the telescope as if it were using the remote. This solution has the merit of simplicity. Most modern mounts also have a serial port (RS-232), which allows for a higher level communication between the computer and the telescope. For example, the computer could order the mount to point at a different object. The mount can also reply to the queries of the computer, for example tell it what are the coordinates it is currently pointing at. For the communication to work, the two systems (computer and telescope) must use the same protocol. To add one more complication, there are many protocols available. The most common is the protocol LX200, named after the mount manufactured by Meade. Therefore, you have to look for the possible connection and protocols available for your mount, then for the protocols that your guiding software can understand – and the latter must also recognize the guiding camera. Of course, the same autoguiding software must be able to work with the guiding camera and the telescope.

However, the autoguiding software and the acquisition software are not necessarily the same, because these two functions are completely distinct.

> To better separate these functions, one can even use two different computers, one for the acquisition of spectra, and the other for pointing and autoguiding. This solution also enables having two separate screens, which I appreciate during my observations.

To verify that your autoguiding software communicates properly with the mount, it is generally sufficient to rapidly move the telescope under computer control.

Some users choose to also connect their mount to the sky map, either to visualize where the telescope points directly on the map, or to select the target to point from the map – most software sky maps allow this. However, in this case there are two pieces of software – the autoguiding and the sky map – sending orders to the mount. This can lead to conflicts. The classic one is to ask to point to a new object, without deactivating the autoguiding. This cohabitation of several programs introduces an extra difficulty: I strongly recommend going step by step, and not try to make everything work together at your first attempt.

There is a third difficulty in obtaining decent autoguiding: the setup of the correction parameters. The autoguiding is an iterative procedure: it constantly loops between the measurement of position, calculation of the deviation and correction of the telescope position. Any iterative procedure is always based on parameters dedicated to the system to be corrected. This means they might be very different depending on your equipment. There is an inevitable learning curve, and things might also vary depending on the atmospheric conditions – for example, the wind might significantly perturb the iterative procedure and require an adaptation of the parameters. There is no perfect solution: it is always a compromise between the speed and precision of the iterative procedure.

The main parameters of autoguiding are, in general, the following: the threshold for the computer to order a correction to the mount, the amplitude of the correction, the direction of the connection. Some systems even allow averaging over several images before measuring the position. This is very useful in windy conditions, for instance...and most systems offer a learning mode, which allows for a quick calibration of the amplitude of the corrections. But be careful: the learning mode is usually very simple – do not hesitate to tweak the parameters as you gain more experience.

> 💡 On an equatorial mount, when you reverse it (passage through the meridian), you need to invert the direction of the autoguiding in right ascension (RA)!

Let me raise another technical point regarding loose mounts. I am referring here to mechanical looseness, especially in the mount drive gears (also called backlash). This looseness is extremely bad for guiding, and it is important to minimize it as much as possible (which might be tricky). In fact, the looseness perturbs only the tracking in declination. The tracking in right ascension requires a constant motion in the same direction. Therefore, the mount is always lying on the same side of the loose part for the RA. Modern mounts can generally compensate for the looseness when moving – but we require such a high precision that this compensation can actually hurt more. A trick to not be too penalized by the looseness during guiding is to misalign the telescope *very slightly* from the Earth rotation axis, so that the mount requires a regular – but not too frequent! – correction in declination. Doing so will force the mount to always be on the same side of the looseness, like for the RA.

Autoguiding can feel like a magic system, which not only can compensate for fine problems (mechanical deformations, etc.), but also larger ones, like the alignment of the telescope. This is only very partially true. Autoguiding can compensate for *some* alignment problems, but this requires the system to "keep up with the star", and it will necessarily affect the quality of the tracking. Keep in mind that the autoguiding is there only to complement a proper setup, and make *small* corrections to allow for very long exposure times. It should not be used to compensate for a poor setup.

Chapter 12

Installing the Spectroscope on the Telescope

By now, you have a telescope you've mastered (you have no problems pointing to a faint star and centering it in the field of your camera with pixel precision), and you have taken your first solar spectrum (complete data acquisition and reduction). The major part is already done !

The rest of the adventure is "just" putting together the two instruments, and to use the telescope to send the light from the star you wish to observe to the slit of the spectroscope.

Keep this picture in mind: the telescope and the guiding stage of the spectroscope serve the purpose of collecting the light from the star and sending it to the spectroscope slit. The spectroscope, in turn, decomposes this light and projects the spectrum on the CCD of the main camera.

When starting, take care to be comfortable: plan to have a table and a chair close to the telescope, from which you will control the whole instrumental ensemble (fig. 12.1).

All the following operations must be done during daylight.

12.1 Stiff Mechanical Match

Attach the spectroscope solidly, with its two cameras (main and guiding) on the telescope. Make sure the position of the slit (which must be in the focal plane of the telescope) is compatible with the telescope output. If you use a focal reducer, make sure it is properly adapted (a reducer is supposed to work only at a specific position of the focal plane).

Set up all the cables for the mount and the different cameras. The selfguiding is not needed in the first acquisitions – you can do the guiding manually. The connection to the computer is therefore not required at this stage.

Once you have done a proper alignment of your telescope, you might be afraid to alter it when mounting the spectroscope. My experience is that

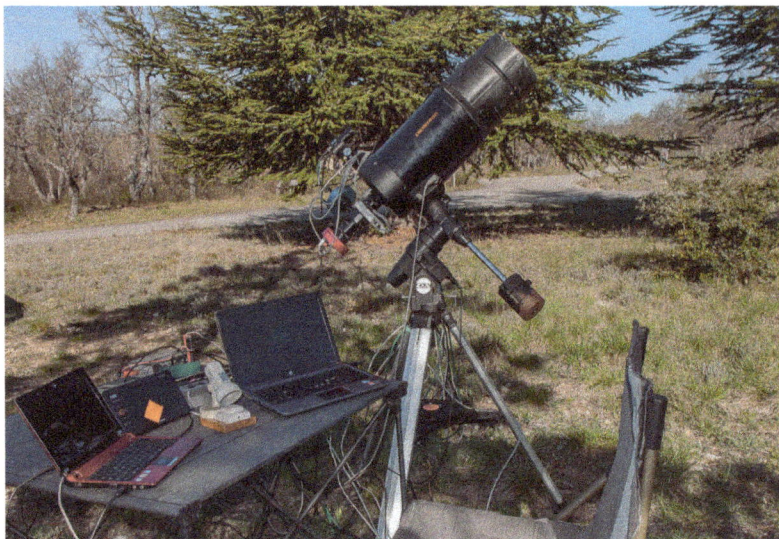

FIG. 12.1 – Photo of the complete installation.

things usually work well – provided the ground on which you have installed the telescope is not too soft. Of course, if you have never tried before, it is better to try to mount the spectroscope on the telescope before the alignment, to avoid problems.

12.2   Orientation of the Spectroscope

Make sure you properly orient the spectroscope on the telescope. What does "properly" mean? It means the slit (which is part of the spectroscope) is oriented so the the motions of the telescope (along the RA or DEC axes) are natural – i.e. along and perpendicular to the slit.

If you have mounted the guiding camera to see the slit horizontally (or vertically) in the guiding image, then a displacement along RA must correspond to a vertical (or horizontal) motion of the stars in the guiding field.

It is a small mental exercise at the beginning, but you quickly get used to it: the key is to take the time to understand the various motions of the telescope and their results on the (guiding) screen when you first mount the spectroscope.

12.3   Balancing and Cables Management

Once as you have mounted the spectroscope, check the mount balance on both axes (fig. 12.2). Since the spectroscope weight is substantial, you might

FIG. 12.2 – Balancing the mount.

find yourself at the limit of the balance (at the end of the counter weight bar with insufficient weight – in the case of a German mount). When this happens, I add weights at the end of the bar. I use "soft" weights that you can buy in any sports shop (those that you can strap on your ankles or wrists). Of course, you need to make sure the total weight of the installation is within the capabilities of the mount.

A step often overlooked by beginners is the "management of the cables" (fig. 12.3). There are many cables from the telescope itself, for the acquisition, guiding, and pointing cameras, the calibration lamp(s), etc. I am convinced this is one of the elements that make spectroscopy "feel" complex. You absolutely want to avoid cables being unplugged in the middle of the night because of the telescope motion, electric cable perturbing the follow-up, hanging cables disconnecting a camera, etc.

These issues are easily solved by tying the cables together, then to the telescope, and finally to the mount. For this, I use small cable ties or strings. Of course, you need to leave enough cable length for all the telescope maneuvers (including the meridian flip), and to allow it to point in all the directions. However, the cables weight must not be supported by the ground, so that the telescope always carries the same total weight, and the balance does not

FIG. 12.3 – Cable management.

change. Since the day I left enough cable length to give the telescope freedom of movement, I definitively solved the cable issues.

Obviously you need cables sufficiently long to accomplish what I outlined above. Investing in sufficiently long cables (in my case I need 5 meters to go comfortably from the telescope to the table I set up next to it) is not a waste of money.

Once you have properly set up the cables, quickly verify that the balance is still all right, and compensate if needed.

12.4 Plugging in

All is ready to start the installation. Plug in the acquisition camera, then the computer. Start the camera cooling and the acquisition software.

Point the telescope to the sky (not directly at the Sun). Do a first acquisition and verify that you obtain the solar spectrum. It must be crisp – if not check the spectroscope focus (see section 8.3). At this stage the telescope does not need to be in focus.

While you are at it, make sure you can take a flat image (white light) as well as a calibration spectrum. Depending on your instrument, the lamps for flat and calibration are either integrated in the spectroscope (refer to the instrument manual), or you need to use external lamps. Actually these images need to be taken by night, to avoid contamination from the Sun, but for now make sure that everything is working.

To take flats you need a bright light source (halogen lamp) in front of the telescope. You might want to put a sheet of white paper or white plastic to diffuse the light and avoid local optical effects, mimicking a light beam coming from infinity (fig. 12.4).

As for all other images, you have to set the exposure time: avoid saturation and make sure the signal is close to 80% of the camera maximum ADU level. Depending on your equipment and the light source, this will require from a few seconds to several minutes.

For the calibration, proceed in the same way, but now it is likely that your calibration source is faint – increase the exposure time appropriately. In this case, however, you don't need to approach levels close to saturation. You need enough signal to measure the position of the emission lines precisely.

12.5 Focus, Guiding, and Telescope

Now, plug in the guiding camera and start the guiding software. Take some images (with very short exposure times, since there should still be daylight), and verify that you see the spectroscope slit sharply and horizontally in the center of the image.

FIG. 12.4 – Diffusing element at the telescope entry.

I believe it is useful to spend some time on the focus. Let me emphasize that all focus settings must be done with a 1×1 binning: this guarantees the best evaluation of the image on the screen.

Three elements need to be in focus on the spectroscope slit:

– the spectroscope itself, since its job is to image the slit;

– the guiding camera, which checks the spectroscope entry (i.e. the slit);

– the telescope, which needs to concentrate the maximum amount of light at the spectroscope entry.

For sure, the slit is the heart of the whole instrument. Put the spectroscope in focus first, its settings are independent from the rest of the instrument. In contrast, the focus of the guiding camera and the telescope are strongly entangled, and there is a trap to avoid. Figure 12.5 shows a scheme of the three elements that need to have the slit in focus

In practice, since you cannot put an eyepiece behind the telescope anymore, the only way to put the telescope in focus, is to look at the guiding image. But beware, if this one is not perfectly focused on the slit, you might think that the telescope is in focus (if the image of the stars is crisp on the guiding image), while it is not (fig. 12.6).

The way to avoid this problem is to always focus the guiding camera on the slit *before* focusing the telescope. What happens if you don't do things in this order? The bottom line is that the star is not in focus on the slit (while it is sharp in your guiding image), and only a small fraction of its light enters in

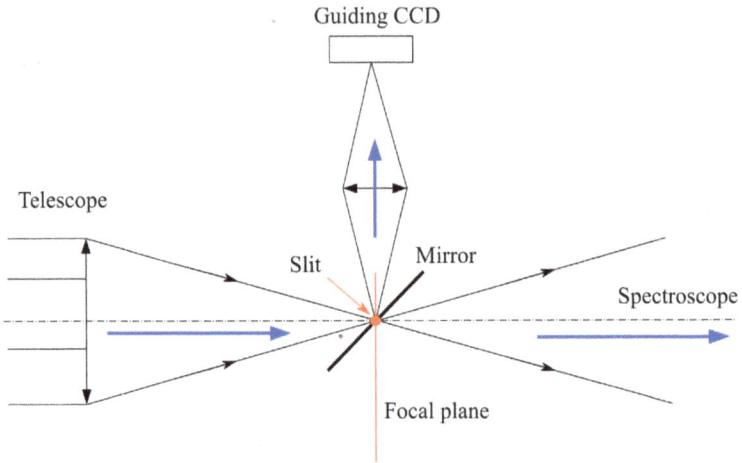

FIG. 12.5 – Three elements focusing on the slit.

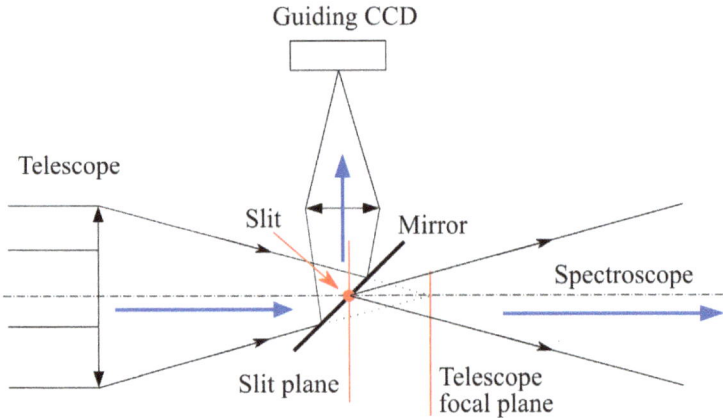

FIG. 12.6 – Bad focus of the guiding camera.

the spectroscope: an exposure which should last only a few seconds can exceed several minutes! With experience, you will be able to understand if there is a problem, since you'll know what the exposure time should approximately be for a given magnitude of the star. But at the beginning, you don't necessarily know this, and it is possible to make mistakes.

Once the guiding camera is properly in focus, don't touch it anymore (and fix its position), and move on to the telescope focus. For this, point a frame on the horizon (a mountain, a pole, ...) and set the focus using the guiding image. Do not waste too much time on this step: you will need to re-do it at the beginning of the night, using stars. The important thing is to be close to the optimal focus, so that the stars will immediately show up on the guiding image.

12.6   Last Checks

Take advantage of the remaining daylight to note the coordinates of the center of the slit (X and Y coordinates, with pixel precision): when it will be dark, you will not see the slit in the image anymore, so it will be useful to know its effective position. If you use an auto-guiding software which allows for displaying a graticule in a given position (like AudeLA, PHD2, or AstroArt), put it in the center of the slit.

Verify that the finder is aligned using an object on the horizon (pole, tree, ...).

When the all the images look right, switch on the mount. If you have chosen to work with the auto-guiding, connect the PC and verify that you are able to maneuver the telescope from it.

The last "critical" step is to make all the equipment (two, or possibly three cameras, and the telescope) work together at the same time. Normally, everything should work fine if you have tested each element separately. But if there are compatibility issues, of the PC is not powerful enough, only now will it become evident. Note that an insufficient power supply can also create problems at this stage.

Let your installation run for several minutes – and if everything works properly, you can rest until the twilight.

While you wait for the night, take the time to choose your first target. Pick a bright one (magnitude less than 4), high in the sky, not too far from the meridian, and if possible, take a hot star (spectral type O, B, or A).

12.7   At the Beginning of the Night...

Now, everything is ready for the acquisition of your first stellar spectrum: it is the "big moment" of the journey we have taken from the start of this book. I will spend some more time on some of the critical points of this essential phase, since it only takes a little effort to make the difference between a mediocre and a good observation.

Verify once again that the installation all works and make sure that the acquisition camera is cooled.

Initialize the Mount on a Star

Trigger continuous acquisitions, and initialize the mount (by pointing at a bright star, known by the mount). You can do this first pointing manually (loosening momentarily the brakes) or with the remote, as you prefer. Use the finder, and put the star right in the middle of your image. If everything is properly set up, you should see the star on the guiding image. At this stage, it is useful to work in pairs: one person controls the telescope, and the other checks the screen.

I'll suggest a trick for when, regardless of the finder alignment, you still don't see the star in the field of the guiding image. Since you are pointing at a bright star, you have a significant flux available. When the telescope is out of focus, the star forms a disk, and it can be bigger than the guiding image. The misalignment between the finder and the telescope should be small (if you did things right!), and therefore the star should be close to the image plane. By progressively de-focusing the telescope, you will see the disk of the star appearing into the guiding image. You can then re-center it, and re-set the focus, little by little. For this to work properly, make sure the visualization threshold of the guiding image is low (so that any image, even faint, can easily be seen), and increase slightly the exposure time (1 or 2 seconds), to guarantee enough counts to make the disk visible.

When the reference star is in the center of the guiding image, tell the mount it is pointing at that precise star. The mount is now initialized (it knows which direction it points at), and can point to any other position in the sky[41].

From now on, you must not loosen the brakes: any manipulation of the telescope must happen through the remote controller or the computer, so that the mount does not lose its reference[42].

Point the Target Star

Now, using the « GoTo » function, point the telescope towards the star you have decided to observe. Continue to take continuous exposure with the guiding camera. Do not activate the selfguiding yet (I strongly recommend to do the first guiding manually).

You should see the star in the guiding image – adapt the exposure time and/or the guiding camera gain to avoid saturation. If you don't see the star, look for it with the finder.

Do a fine focus of the telescope, so that the star is as point-like as possible in the guiding image. Beware: at this stage, only modify the telescope focus, not the guiding camera focus (you would not see the slit anymore, and almost no light would enter in the spectroscope – see above).

To quantify the focus precision, you can measure the full width half maximum (FWHM – AudeLA and other software allows you to measure this quantity on the fly). The effective size of the star in the guiding image strongly

---

[41] If you have a mount with absolute encoders, you don't need to do this initialization.

[42] Again, if your mount have absolute encoders, then you can move it freely - the actual position will never be lost.

depends on the focus of the telescope and the observing conditions (*seeing*), but generally you should aim for 2 to 5 pixels.

> Make sure the alignment allows for tracking of a few minutes without trouble. This is necessary for your first spectral acquisitions. If this is not the case, go back to the previous steps (polar alignment, management of the cables, etc.)

If you have carefully taken note of the position of the slit on the guiding image, put the star there, with slow movements of the mount. If you don't know where the slit is on the image, you can always shine light on the telescope entry with a pocket lamp, or directly on the slit with the calibration lamp (if integrated in your spectroscope): the slit will appear clearly.

In the end, the star should be in focus, non-saturated, visible on the screen and close to the center of the slit. There are many parameters to deal with at the same time – take the time to set each one of them. It is a critical phase, since we work at the core of the union between the telescope and the spectroscope, and we are sending the light of the star in the spectroscope. Start with a bright star, to obtain your first spectrum more easily, even if everything is not perfectly set. This is the crucial step of sending the stellar light into the spectroscope that determines the quality of your observation. The better the settings at this stage, the higher the intensity in the spectrum: the precision of the focus and the alignment on the slit determine the quality of the spectrum.

Center the Star on the Slit...

It is not enough to put the star in the guiding field: you need to put it in the center of the spectroscope slit. Use slow movements of the telescope to precisely position the star there. You might think it is enough to put it anywhere on the slit, but it is a good habit to work always with the same position of the star on the slit (so that the spectrum is always in the same position on the main image), close to the center, since it is closer to the optical axis of the telescope.

When the star crosses the slit, you must clearly see its luminosity go down. It is very important: if the intensity of the light decreases, it means the light is going into the spectroscope.

To make sure you can clearly see this dip in luminosity, you need to choose the exposure time (possibly the gain) properly and correctly set the visualization thresholds of the guiding image (make sure you deactivate the automatic calculation of the threshold). A star never completely disappears, since its image is always larger than the slit. But if the ensemble of your installation is properly set, you should see the star dim when it crosses the slit.

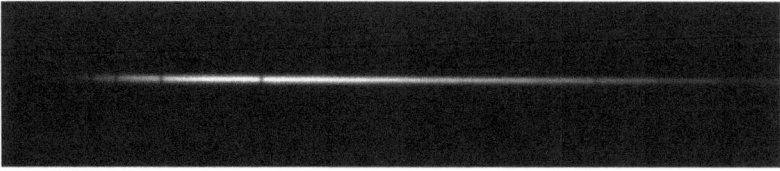

FIG. 12.7 – Stellar spectrum.

... and Keep it in the Slit

You now have to keep the star constantly in the slit for the whole duration of the exosure, which can last a few seconds...to a few hours! Once again, the precision required is too high to rely uniquely on the quality of the mount tracking.

When you observe a bright star, the exposure time is short (order of seconds), and the tracking problem does not exist. But as soon as you observe stars requiring several seconds of exposure, it is likely that you will need to adjust the position of the star on the slit during the exposure.

I suggest that you do the correction manually at the beginning – called manual guiding. This means that you need to constantly observe the star in the slit, and as soon as you see it drifting away, compensate using the mount remote, using the slowest speed. This gives you a feeling for the telescope motion and the required corrections.

You will quickly realize that this is constraining, and that it is much better to use the autoguiding (see section 11.5). Moreover, the autoguiding ensures a reproducible and better quality tracking.

Keep in mind that if the tracking is not optimal – either because you are distracted, or because the autoguiding is not properly initialized – the outcome will be a spectrum with lower signal (and thus worse signal to noise ratio – SNR), but with the same intrinsic quality (resolution).

Take a Spectrum

Once the star is well centered on the slit, its light enters in the spectroscope and forms a spectrum on the acquisition CCD. Thus, it becomes possible to image it. Since you chose a bright star, a short exposure time (order of seconds) will suffice. Take a first spectrum and have a look at the result.

Now, the spectrum is a simple horizontal line in the middle of the image (fig. 12.7), while for the solar spectrum we had a broad band. This is because the Sun is an extended, resolved object from Earth.

Improve the visibility of the spectrum on the image using the visualization thresholds. You should immediately see some absorption lines. This first spectrum is important: you have now managed to connect all the elements together, and you will be able to progress quickly.

Get in the habit of checking the maximum intensity level of your spectrum. If it is small compared to the camera limit, increase the exposure time appropriately. For example, if the level is 5,000 ADU, and your camera has a upper level dynamic ADU value of 65,536 ADU (16 bits camera), you can increase your exposure time by a factor of 10. Conversely, if the image is saturated, take another with a shorter exposure time, until it is not saturated.

> When you chose the exposure time, aim roughly at 70-80% of the camera's upper ADU value. Why not more? Because it is likely that between two exposures the maximum level reached changes, and you absolutely want to avoid saturation. Thus, it is wise to save some buffer below the saturation level.

Now take several images of spectra in a row. You can see variations between the spectra:

- if the spectrum moves vertically from one image to the other, it means the star moves along the slit. Try to make it stay in the exact same position;

- if the spectrum is wider/thinner from one image to the other, or if it is split, it means the star moves along the slit *during the exposure*[43]

- if the maximum intensity is the not the same in all images, it means that the amount of light entering the instrument varies from one acquisition to the other (see below).

More Light

I take the risk of being pedantic: it is essential to maximize the amount of light entering in the spectroscope. This is what determines the quality of your spectrum. I insist because it is very easy to lose 90% of the light before the slit, but not so easy to notice it. There are at least three ways to loose light:

- the telescope is out of focus, so the image of the star is not a point similar size as the slit, but a disk. In this case, only a portion of the light from the star enters the slit (fig. 12.8);

- if the star is not positioned exactly on the slit, the result is the same: only a fraction of the stellar light enters in the slit (fig. 12.9);

---

[43] Beware: if the spectrum is split on all images, there might be an issue with the focus. With Schmidt-Cassegrain type telescopes (using a central blocking element), you can see the obscuration by the secondary mirror in the middle of the spectrum.

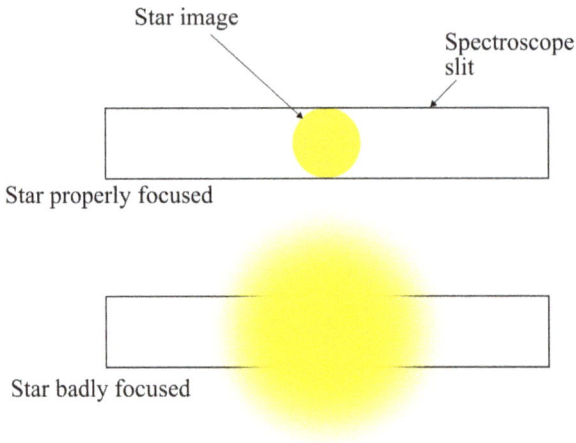

FIG. 12.8 – Focusing the star on the slit.

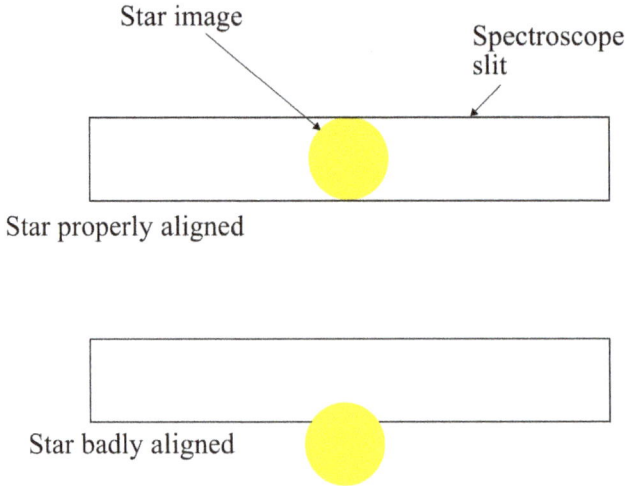

FIG. 12.9 – Aligning the star on the slit.

– issues with the tracking, seeing, or wind can also cause significant light losses – with the same result. Seeing is hard to deal with, and so is the wind if your instrument cannot be protected with a dome or shelter. On the other hand, the tracking can always be improved: with a little practice, you can see on the guiding image the main cause of light losses (oscillation, shift, ...) – and compensate for it.

If you can easily remove the slit from your instrument, you can do an extreme experiment: compare the total flux measured in your spectrum with and without the slit. If all the light from the star crosses the slit, the results should be the same – this is never true in reality. If you obtain 70% of the flux with the slit, it is a good result (but this can change a lot depending on the conditions).

Now, you have accomplished the hardest part: you know how to produce a stellar spectrum. Take the time to do this operation many times, with many different stars. The only thing left is to do complete observations – since the raw spectra cannot provide much information without the reference images necessary for the data reduction.

Chapter 13

Spectroscopic Observation of another Star

We now have all the elements to make a complete observation, from the data acquisition to their reduction. In this chapter, I will only repeat what I already described. The new element is how they all link together.

Start by choosing the star(s) you want to observe. Once again, chose a bright star high in the sky; there will be time to try to obtain a spectra from fainter stars later.

Choose the reference star to be used to evaluate the response curve of your equipment. You can refer to section 9.8 for some catalogs of reference stars. The reference star should satisfy the following criteria:

- it should not be too bright, to allow exposure time longer than 30 seconds (thus reducing the possible chromatic aberration from the slit). For example, aim for a magnitude of about 5;

- it should be close to your main target (same height above the horizon and possibly at less than 5 degrees from it);

- it should be a hot, B-type or A-type, to get a smooth spectral energy distribution;

- you need to have a reference spectrum of this star (e.g. in the ISIS database);

- it should be high enough in the sky (more than 30 degrees), to avoid most of the chromatic effect from the atmosphere.

13.1 Starting the Observation

Let me repeat briefly the start-up process.

Start your equipment when there is still daylight. Turn on the PC, verify that it connects properly with all the cameras and the mount, and take a few

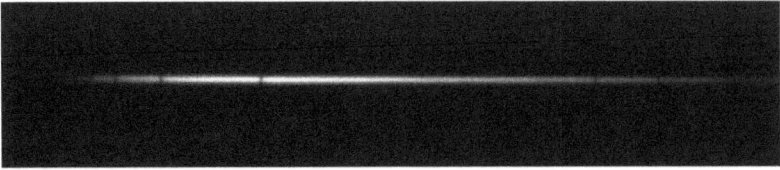

FIG. 13.1 – 2D image: spectrum of a hot star.

test images – for the spectroscope, the guiding camera, and, if possible, for the pointing. Turn on the cooling of the acquisition camera.

Take note of the exact position (with pixel precision) of the center of the slit on the guiding image, and put a grid on it with your software, if possible.

13.2 The Reference Star

Point to the reference star, and make sure it is right in the center of the slit. Take a few acquisition images to find the optimal exposure time – probably not more than a few seconds if the star is bright. The intensity of the spectrum should be about 80% of the camera's upper ADU value. You should see a continuous spectrum, with very prominent (hydrogen) Balmer lines (fig. 13.1).

Then, switch on the autoguiding, and take a series of images with this exposure time – say, 15 images (you can afford to use a large number, since the exposure time is short).

13.3 Point to the Target Star

Then, move on to your target star. Proceed in the same way: verify that it is in the middle of the slit, chose the optimal exposure time – which is probably different from the previous value – switch on the autoguiding and take a series of spectra (about 15).

13.4 Take all the Reference Images

After taking the first raw spectra, take all the reference images which will allow you to do a clean and proper data reduction.

- Take seven (or more) bias frames (telescope closed and zero exposure time). These images can then be reused for several nights.

- Take seven (or more) flat images. Use an incandescent (e.g. tungsten) lamp (continuous spectrum). Choose the exposure time to have a maximum level of about 80% of the camera upper ADU value. These images can be reused for the entire observing night, but it is preferable to take new ones each night, to account for changes in the dust along the optical path.

- Take one (or more) spectra of a wavelength calibration source. Once again, choose the exposure time to have a maximum level of about 80% of the camera upper ADU value. You should take a new calibration spectrum regularly (how often depends on the precision you need for the wavelength calibration) to account for possible variations in the instrument (thermal effects, deformations, etc.).

- Take seven (or more) dark images. The exposure time should be at least as long as the longest exposure time of the whole series of images you already have (including the reference star, target star, flat, bias, and calibration). These can be reused for different nights – provided the CCD is cooled to the same exact temperature each night.

Once you have all these images, you can shut down your instrument setup – or take the opportunity to observe some other stars, maybe of different spectral type (stay with bright sources at the beginning).

13.5 Data Reduction

Your precious images are now « in the box » - congratulations! You now have all that is necessary for a proper data reduction.

The data reduction sequence is exactly the same as the one used for the solar spectrum, although you now have a reliable (and external) source for the calibration, and the observation of the reference star allows you to do a precise, independent determination of the instrumental response.

Once again, I rely here on the sequence of operations in the ISIS software, but you can perform the data reduction with the software of your choice.

Start by creating the master bias, dark, and flat images.

Then, reduce the data of the reference star. Go to the tab « 1 - Images », and load the first raw image of the series. Unlike the observation of the Sun, the spectrum is now a thin line (since the star is a point-like source), and you can use the spectrum of the sky surrounding the stellar spectrum to eliminate the sky background. In ISIS, make sure the region for the calculation is adjusted to your image, and uncheck the box « Sky background not removed » in the tab « 2 - General ».

> 💡 To correctly place the region for the calculation of the spectrum just double click on the spectrum to center it. If needed, you can make fine adjustments in the tab « 3 - Calibration ».

> 💡 If the position of the star changed during the acquisition, make sure all the images of the series fit in the same region for the calculation. For this, use the left and right arrows (next to "Display"), which allow rapid switching between the images of a series.

In the tab « 2 - General », indicate the name of the target and the parameters of the observation (instrument, site of the observation, and observer in the bottom right of the window). These elements will be saved in the FITS file header containing the results of the data reduction.

Then, click on « Next » to go to the « 3 - Calibration » tab.

For wavelength calibration, find between five to ten characteristic spectral lines of known wavelength. Chose these lines so that they are spread over the whole spectral domain covered. Calculate the dispersion law (polynomial of third or fourth degree), and save it.

> 💡 If you have the calibration module for Alpy, ISIS can automatically calculate the dispersion law – you only need to provide the position of one of the easiest lines to recognize (using the mode « Alpy 600 (with calibration module) » in the "Settings" menu).

Once your spectrum is calibrated in wavelength, you can compare it to other spectra – and in particular you can use it to calculate the instrumental response. Proceed as for the solar spectrum: take the ratio of the spectral profile obtained to the theoretical spectrum, and then smooth the result (moderately) to get rid of the deep lines in the curve. Save this spectral instrumental response with a simple name.

Then, go back to the tab « 2 - General » of ISIS, and fill in the field « Instr. responsivity » with the name you just used. You can now run the data reduction pipeline again: this time all the operations are done at once, and in few seconds you get a calibrated spectrum corrected for the instrumental response.

Doing the data reduction of your main target is now easy: you already have the master images, the dispersion law, and the instrumental response... Actually, you almost have nothing else to load into ISIS, except the raw images!

To do this, go to the first tab (« 1 - Images »), and load the first series of images of your target. Verify that the region for the calculations is unchanged (for the entire series). Then, click on « Next », and only indicate the name of the target (which is necessarily different from the name of the target star). All the rest is automatic: click twice on « Next » to reach the tab « 4 - Go », and start the calculation by clicking on « GO ». After a few seconds, the calculation is done and you can see the reduced spectral profile: you get a calibrated spectrum corrected for the instrumental response.

The rapidity of the process to obtain the spectrum shows all the power of ISIS: it really is a tool which allows for productivity, since you just need to spend some time on the parameters for the spectrum of the reference star, and then all of them are saved to be used with the next stars.

ISIS takes the (good) initiative to save the obtained profile with different formats (FITS, DAT). It also saves the results of the calculations with the .log extension. It also allows one to create a PNG file with Gnuplot (if installed on your computer).

With a bit of practice, the first reduction takes a few minutes, and the following only take a few seconds: it is possible to do the data reduction « in real time », i.e. during the acquisitions for a star, you reduce the data of the previous one.

13.6 Going Beyond

I have said at the beginning of this book that I am a follower of the notion of continuous improvement. Once you have taken your first stellar spectrum, you have done the series of operations that led to the result: now it's time to switch to the « continuous improvement mode ». That is, you now know how to get a spectrum, and you can improve the result step by step.

Obviously, you will regularly redo this operation, and each time you will try to spot the « tricky points » you encounter. Take the time to analyze the cause of these tricky points, and modify your procedure to improve it.

Each situation is specific, but let me outline a few paths to quickly improve your observations.

- Activate autoguiding. Force yourself to always put the star in the middle of the slit, so that it is always in the same position in the raw images. Autoguiding has the advantage of maintaining the same position in both directions (X and Y). The vertical alignment (across the slit) is necessary to get more of the light into the spectroscope, and the photometric stability. It is less critical if the star moves horizontally (along the slit)... but the result is a vertical displacement of the spectrum in the raw images (I suggest making some trials with the star in different positions along the slit, to see the effect on the spectrum). Since the position of the spectrum is one of the parameters considered in the data reduction, you will save time by having the star always in the same position in different series.

- Depending on the kind of spectroscope you use (especially if you are working at high resolution), you might face problems because of deformation of the instrument, which can impact wavelength calibration. To prevent this, you should take calibration spectra after each observation of a star (without moving the telescope, so that the calibration spectrum is taken in the same position as the stellar spectrum).

– you need dark images (or just *darks*) for the data reduction. This takes time, since the exposure time needs to be at least as long as the longest exposure of the night. But they only depend on the camera and its temperature, not on the rest of the instrument or the observing conditions. Thus, it is not necessary to take new ones each time! The same applies to the bias images – although these are faster to take. You can even produce a « library of darks and biases », i.e. use the cloudy nights to make a very large series of darks at different temperatures and with different exposure times. You can create master images immediately: this is all time you save during future nights. Beware: this works only for a specific camera. If you change the camera on your spectroscope, you must not use the darks from another camera.

– The number of spectra of the target object that you need very much depends on its magnitude. For bright objects, saturation is reached in at most a few seconds. In this case a few images (five to ten, and the same for the reference images) will suffice. But you will rapidly notice that with fainter objects (say starting from magnitude 7 – but this of course depends on your telescope), the exposure time increases very fast (several minutes, typically) and it becomes impossible to reach saturation – or even to exploit the full dynamical range. In such cases, it might be better to use several acquisitions of constant exposure time – I often work with exposures of 5 minutes (300 seconds). Since the spectra are digital and the CCD is linear, these can be summed. A one hour observation becomes a series of 12 consecutive images. Collecting many images has several advantages:

  – similarly for the reference images, it allows you to average the noise in the images and add up the signal: this improves the signal to noise ratio (SNR);

  – if an image is not good (guiding problem, cosmic ray, etc...), it doesn't ruin the whole series;

  – using a constant exposure time (300 seconds if you follow my example) simplifies the use of a library of darks, since the exposure time is the same for all images.

– Take notes during your observations. Every professional observatory has a « logbook ». The users (astronomers and support staff) write down in it every event regarding the observations of the instrument worth noting. You will learn a lot by writing your own « logbook ». It can be either a paper notebook, or in digital format. The important thing is to carefully archive it with the data. Take note of everything: program for the night, sequence of observations, etc. This information will be important later, when you try to remember the precise conditions when you took your data.

Chapter 14

Quality of the Spectrum

You have now obtained your first stellar spectra, and with time you will become very fmailiar with all the procedures– until you can do them in a "routine mode".

Soon, a crucial question will arise: are my spectra good? It should be a constant question, since you are never safe from errors, big or small.

This is a short list of the most common defects found in stellar spectra:

- the spectrum is noisy (bad SNR). Typically, this is caused by excessive light losses at the slit (see section 12.7). You could increase the exposure time, but make sure you have optimized the light injection first;

- problems in wavelength calibration. It is easy to detect (see section 14.3);

- problems in the continuum correction: the overall profile of the spectrum does not look as expected;

- problems in the continuum level: typically it indicates a bad pre-processing (bad dark or bias correction);

- ... and it happens regularly that the spectrum does not look at all as expected: maybe you got the wrong star!

In this chapter, I will outline some ways to evaluate or measure the quality of your spectra, and avoid as many as possible of the problems listed above.

To do these sanity checks, I often use the software VisualSpec, developed by Valérie Desnoux[44]. It allows one to compare two spectra, embeds a library of chemical elements, and is capable of performing many types of measurements on the profile. It is simple – and free – software.

---

[44] VisualSpec : http://www.astrosurf.com/vdesnoux/

14.1   Read the Outcome of the Calculation

The first thing to check after data reduction is to look at the result log given by the reduction software. In ISIS, there is a window showing the details of the calculations, including intermediate results (the so-called "log"). This information can reveal at a glance an anomaly in the process (bad spectrum in the series, calibration problem, wide diversity in the spectra, etc.). I will not go into details about the calculation log here, but I suggest you have a careful look at it.

14.2   Compare with other Observers

It is by far the most efficient method to make progress: compare your spectra to those obtained by other observers. More and more free access databases of spectra (in FITS format) are available, and you can easily download them and use them for comparison. Let me cite again the BeSS database[45], which is suitable for this exercise. You can download spectra taken with different instruments for all bright Be stars, and you will certainly find one roughly corresponding to your setup.

This is the advantage of having a standard format for the files: if you save your spectra in "FITS" format, you can instantly compare them to each other, even if the resolution or the spectral domain are not exactly the same.

When you carry out such comparison, you can check for several key points:

– did you observe the right star?

– is the noise level right for your instrument and exposure time?

– Is the continuum and/or instrumental response correction right?

The answers to these questions are the main indicators if there is a problem with the observation, either during acquisition or the data reduction process.

After these checks, you can go further: provide your spectra to the community[46]. By doing so, you will get an "independent" validation of your measurements, and in case of troubles, pertinent advice to improve. It is always a bit scary to expose yourself like this, but the community of amateur spectroscopists is friendly, and you will always be warmly welcomed (everybody recalls their own beginners mistakes!).

My experience tells me that beginners are reluctant in showing their results. Nevertheless, it is really the best way of improving. Also, do not hesitate to ask questions on the various forums: something bugging your mind for days might find an answer in a few minutes if you show it to more experienced eyes.

---

[45] BeSS : http://basebe.obspm.fr/basebe/

[46] For example on the ARAS forum: http://www.spectro-aras.com/forum/viewforum.php?f=10

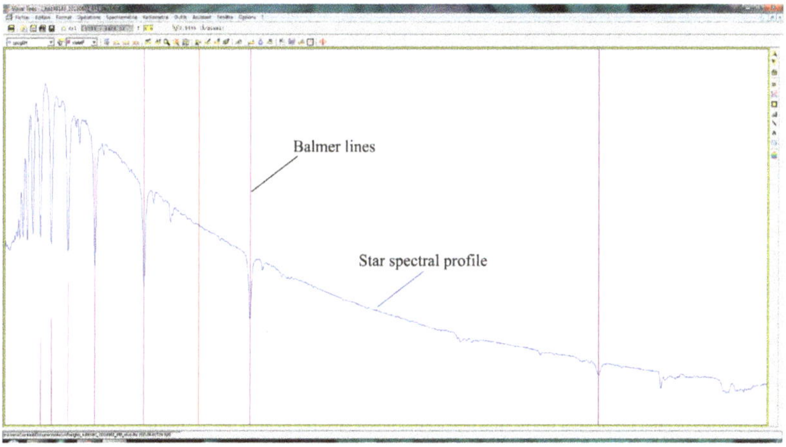

FIG. 14.1 – Screen dump of VisualSpec – Spectrum of *lam Cyg* with Balmer lines overplotted.

Let me cite once again the example of the BeSS database: you can submit your spectra there – they are even awaited impatiently! These spectra are validated by experienced observers, and if they are rejected (it happens), you will get all the advice you need to solve the problems you faced.

14.3 Verify the Wavelength Calibration

Wavelength calibration is one of the critical steps of data reduction, it is always better to check that everything went right there. There are some simple quick checks, depending on your resolution:

– at low resolution, most stars show hydrogen (Balmer) lines, which are usually deep and easily identifiable. It is easy to check that they are in the right place: 656.3 nm for H_α, 486.1 nm for H_β, etc. Using VisualSpec, you can easily show the hydrogen lines: figure 14.1 shows a spectrum of *lam Cyg* perfectly calibrated (all the Balmer lines are exactly in their proper position);

– at high resolution, there are "telluric lines" around H_α. These are the signature absorption lines of humidity in the Earth's atmosphere (from which their name follows), and therefore, they are never significantly Doppler shifted. The depth of these lines strongly depends on your observing conditions (altitude, humidity), but 80% of the time they are easily seen, and it is a very simple way of verifying the calibration

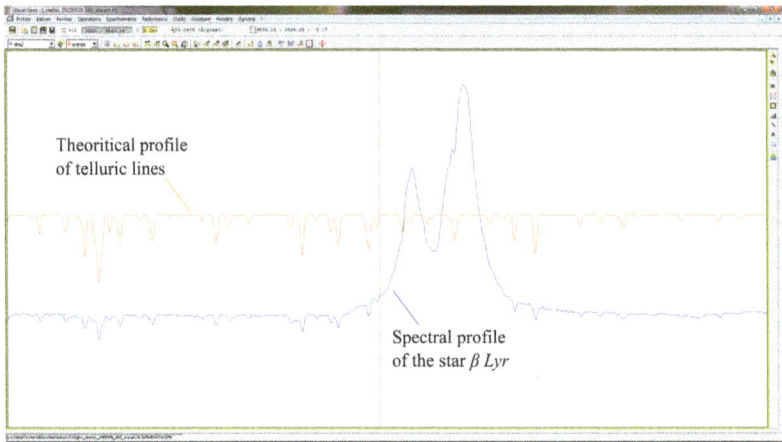

FIG. 14.2 – Screen dump of VisualSpec – spectrum of β *Lyr* with telluric lines overplotted.

of your spectrum. VisualSpec allows one to show them with a click (fig. 14.2).

– in both cases, you can use a reference spectrum of a star of similar spectral type, and the comparison will immediately reveal anomalies.

14.4 Non-uniform Intensities in the Observations

Another simple element to check is the relative intensity of the various raw spectra. In the vast majority of the cases, the light source is extremely stable on a timescale of a few minutes or hours, and an exposure time of x seconds should always yield the same intensity level. If you have significant variations of it among your raw images, it is an indication that there is a problem in your installation. Most of the time, the problem is light injection into the slit (see section 12.7).

This check can also be done using observations taken several days apart with the same instrument. The quality of the sky (turbulence, haze) can have a relevant effect – but roughly speaking, an observation of the same star, with the same instrument and exposure time should yield the same intensity level of the spectrum.

The advantage of this criterion is that it is strongly correlated to the intrinsic quality of the spectrum: if the intensity is not uniform, it means not all spectra are as good as they could be. On the other hand, if all are uniform, it is unlikely that they are all bad (except if the star is not centered in the slit).

14.5 Measurement of the SNR

The SNR measurement is tricky: I am not aware of a universal way of doing it. Measuring the ratio of the intensity of a line to the continuum is relatively easy (using ISIS or VisualSpec, for instance), but measuring the noise level in a spectrum requires some experience.

To do this, you need to remove the continuum (professional astronomers call this *normalization*). Then you need to make sure that no lines perturb the continuum in the region you use for the noise measurement – this is very target dependent.

Nevertheless, it is a very good habit to spot in your spectrum a domain free of lines, and measure the noise shown there. The larger the domain, the more precise the measurement – provided that you have subtracted the continuum level from the spectrum.

If you always use the same measurement technique, you get an objective measurement of the noise, which is a direct indication of the quality of your spectrum.

14.6 Level of the Signal for your Instrument

All I have described above (warnings in the processing log, non-uniform intensity levels, overall level, noise) combine to information you will rapidly know for your setup: for a given magnitude, what is the exposure time you need to get an acceptable level in your spectrum. Of course, the observing conditions (seeing, wind, haze) and the type of object also matter, but roughly speaking you will know what such magnitude requires regarding exposure time.

It is possible to push this reasoning a bit further and make it more objective. In fact, to first order, the level of light L received by each pixel of the camera only depends on a few parameters:

- the effective diameter D of the telescope (light flux captured – possibly corrected for the central obstruction);

- the dispersion of the spectroscope (or, more generally, the spectral domain ρ covered by each pixel);

- the effective exposure time T;

- the camera sensitivity S.

The level L is then proportional to:

$$D^2 \times \rho \times T \times S$$

Consequently, one can calculate a generic level L' comparable for all instruments:

$$L' = \frac{L}{D^2 \times \rho \times T \times S}$$

Therefore, for a given object (assumed to be stable in time), and a given spectral domain, you can compare this value with the one obtained by another observer, with another instrument – even very different from yours. This is another strong indication of the quality of your spectrum.

Chapter 15

Ready for the Adventure

You now have in hand your instrument, and you have mastered the whole process of producing spectra and evaluating their quality. To wrap up, it seems useful to include some final advice on how to make consistent high quality observations – I have already said most of this in the previous chapters. You might want to come back to this chapter from time to time, according to the progress you make during your observations.

15.1 The Typical Observing Session

I summarize here the main elements to carry out a good observing session, i.e. to produce good data in an efficient way. It is a basic checklist that you can use in all your observations.

Preparation – during Daylight

- Prepare the list of targets and define the order of the observations based on their position on the sky. Make also a list of the reference stars associated to each target.

- Start the equipment and cool down the CCD (do not exceed 80% of the maximum nominal power).

- Verify the parameters of the session in your acquisition software: name(s) of the observer(s), location, instrument, working folder, ...

- Set up a folder dedicated to the observing session - Check that spectroscope objective is properly focused.

- Verify the time on your computer (since the date and time of the images are based on it).

- Set up an "observing logbook".

- Quickly try out the acquisition of many kinds of images: sky (solar spectrum), calibration, flat, and make sure everything is ok (crisp images, correct focus, file headers, etc).

- Carefully place the position of the slit and focus the slit on the guiding image.

- Start the acquisition of dark and biases (or find them in your library of darks).

At Twilight

- Verify the alignment & reference position of the mount using a bright star.

- Verify the telescope focus.

- Verify the autoguiding parameters (test for a few minutes).

- Possibly, take a set of reference images (calibration, flat, bias, darks).

During the Observations

For each target (or reference star):

- point at the target, place the target in the slit and activate the autoguiding;

- start the acquisition, taking care of using the right exposure time for the target;

- take the calibration images (if necessary for the precision required);

- fill in the "observing logbook" with any significant information & event which might help you in the future to remember the observing conditions;

- verify that the images are properly saved in the folder for the observing session;

- *(reduce the data as soon as possible)*;

- Deactivate the autoguiding, and start over with the next target.

At the End of the Night

- Verify that you have saved all the images for the data reduction (dark, bias, flat, calibration, reference stars, target).

- Properly archive all the raw images, if possible, before reducing the data.

- Reduce the data.

15.2 Improve your Observations

You have made your first spectra, starting from the acquisition and ending with the data reduction. With practice and experience, you can constantly improve your observations in both directions: the quality of the obtained data on one hand, and productivity on the other one.

15.3 Improve the Quality of the Data

Signal to Noise Ratio (SNR)

The light of the stars is faint... and you spread it with your spectroscope. At the end of the optical path, only a few photons reach each pixel of your CCD. The quality of your spectrum is directly related to the quantity of "Signal" collected by your CCD detector compared to the "Noise" in the instrument, which sets the detection limit (SNR).

You can improve the SNR in two ways: either you increase the signal, or you decrease the noise.

Reducing the noise is difficult. You quickly hit physical limits. But it can be done to some extent. For example, cooling the camera, you reduce the internal activity of the CCD, and consequently the noise.

The best way of increasing the SNR is to increase the signal. This can be achieved in several ways:

- Increase the exposure time (the SNR increases like the square root of the multiplication factor on the exposure time);

- Increase the size of the telescope (this is why astronomers always want bigger telescopes);

- Make sure most of the light enters the spectroscope. Both the image of the star and the slit are very small (a few micrometers), and it is very easy to lose a significant part of the light at the spectroscope entry.

There are three main causes of light losses at the spectroscope entrance:

- The telescope focus. If the focus is not optimal, the size of the image of the star is bigger than the size of the slit, and only a small fraction of the light enters the slit;

- The position of the star in the slit. Placing and maintaining the star exactly in the slit requires precision – and if this is not done properly, you might lose a large portion of light;

- Tracking and/or wind issues, which can be mitigated using the right autoguiding parameters.

A quite simple way to check that most of the light enters the slit is to observe the dimming of the star in the guiding image. When the instrument is well set up, the brightness of the star decreases when the star crosses the slit. To observe this phenomenon clearly, make sure you chose the right exposure time of the guiding camera and the visualization thresholds, to prevent saturation.

> When the target star is dimmer on the guiding image, it means its light is going to the spectroscope.

With experience, you will know your instrument, and you will have an expectation for the signal level of a star of given magnitude. You can also compare your measurements to those of other observers with similar equipment. If the level is below the expected value, take the time to understand what is causing the problem. To improve the SNR even more, you can observe the same star regularly. Choose a well known target to make sure it is stable and measure the intensity of the spectrum over a short range of wavelengths. Since neither the star nor your instrument change, you should always find the same intensity. In practice, you will see variations. Your mission is simple; reduce these variations as much as possible.

Resolution

With a slit spectroscope, the spectral resolution is set intrinsically by the instrument, provided it is properly set up. Your data reduction software probably also calculates the resolving power[47] of your spectrum. Carefully check the value you get. It must be close to the nominal value of your spectroscope. If not, fine tune the focusing of the acquisition camera (refer to the manufacturer documentation)

> The resolution is calculated from the full width half maximum (FWHM) of the lines in the calibration spectrum. The finer these lines are, the higher the resolution.

Wavelength Calibration

Take special care in wavelength calibration. The data reduction software uses a dispersion function to map the pixels to the corresponding wavelength. The function is calculated from the calibration image – either of a spectral

---

[47] The resolving power indicates the capability of your instrument to reveal details in the spectrum: $R = \lambda/\Delta\lambda$, where $\Delta\lambda$ is the smallest detail visible.

lamp showing multiple lines, or a hot star spectrum. This function must be re-evaluated depending on your instrument and the precision required (this happens more often when working at high resolution). In some cases, the dispersion function itself is stable, but there might be slight offsets from one observation to the other, due to small mechanical deformations in the instrument. In such cases, the calibration image is only used to shift the dispersion function. A quick way to verify your dispersion function is to measure on your spectrum the precise position of the (hydrogen) Balmer lines, or the telluric lines from the Earth's atmosphere. These are easily seen in almost all stellar spectra, and since they are well known, it is easy to measure the offset between their theoretical wavelength and the effective value obtained. The precision of the calibration can be quantified using the average value of the differences between these two values over all the available lines (or better the RMS[48]).

Keep in mind that the dispersion function can slightly change when the position of the star changes along the slit. It is always better to put the star precisely in the same position along the slit.

Autoguiding

Successful autoguiding depends on your equipment. It is a crucial element for the quality of your measurements. It is not only a matter of comfort. Autoguiding guarantees better reproducibility of your acquisitions. Its setup is also important. If it is not properly tuned, it is likely the star will oscillate accross the slit – and the amount of light entering the slit will drop. Make some trial runs with different parameters and check the intensity of the spectrum each time.

> To get an optimal setup, make a series of images, and measure the stability of the intensity from one image to the other. If you see variations, probably the autoguiding is oscillating.

Cosmetic File

Every CCD has "hot pixels". These are much more sensitive than the others (it is a defect of the chip). The values obtained from a hot pixel are not reliable, and it is wise to remove them from the image. To do this, your data reduction software will do the following:

- Find the hot pixel using a dark image. (If the camera were perfect, all pixels would have the same value – zero (in absolute darkness). In practice, this is not the case. It is common to find 100 to 500 hot pixels above the normal dark current level in an image);

---

[48] RMS = *Root Mean Square*, the square root of the average of the squares($\sqrt{s^2}$).

– For each hot pixel, the value will be replaced with the average of the neighboring pixels.

Refer to your imaging software documentation to activate the hot pixels correction – it is a simple way of improving your measurements.

Instrumental Response

We covered the instrumental response in section 13.5. The easiest way of obtaining a response curve to correct for this is to observe a well known star and divide the obtained spectrum by the theoretical profile. In reality, the question of instrumental response is complex, and if you need accurate spectra, you need some extra care. For example, the reference star must be similar to your target – the same magnitude, the same weather conditions, and the same altitude in the sky (this is obviously impossible in practice). The instrumental response depends on all these parameters. You also need a reference star which is unaffected by "reddening" from interstellar dust. A bad instrumental response curve directly impacts the general shape of your profile.

Obtaining a good instrumental response curve is a matter of experience. Do not hesitate to do trials and compare them to each other.

Organization and Archive of your Data

Your spectroscopic observations will rapidly result in a large amount of data. To properly reduce these data, and find them later, you must properly organize them. For this, I strongly suggest creating a folder for each observing session (a session is for me one single night of observations). Put your "observing logbook" in this folder.

This is a very important habit. Archive all your data, especially the raw images. It is possible that in 10 years you will want to reduce these images with new tools or a reseracher may want to reduce these images with his/her own tools. Remember that your observations are most likely unique – it would be a pity to waste all this work, which might turn out to be a treasure in the future. Thus, take special care in classifying and archiving your data.

> 💡 I have a simple organization scheme. For each new observing session (ie night), I create a new folder, and name it with the date and the main theme of the observation. I put my "logbook" in this directory, and create three subfolders: "raw images", "pre-processing", and "results". By the end of the night – possibly before the data reduction – I save all raw images in the first subfolder. At the end of the data reduction process, I save all results and intermediate images in the corresponding subfolders.

15.4 Improve your Productivity

Keep the same Configuration

The best way to improve the productivity of your observations is to always use the same equipment and observing protocols. Of course, you might be tempted to improve your equipment (it's actually a rather good thing), but each time you modify a piece of your equipment, you need to make the appropriate changes to your procedures and gain an understanding of the new configuration. Sometimes, it is better to stabilize the equipment, to ensure quality and productivity.

Library of Darks

Darks take a long time to make (the exposure time is at least as long as the longest exposure for a stellar spectrum). The more dark images you make, the better (they allow for a significant improvement of the SNR). Therefore, they are time consuming – and if you take them at the end of the night (once you know the longest exposure time), you will easily get bored.

However, darks only depend on the camera. For a given temperature and exposure time, the dark image is always the same – similarly the read out noise. They do not depend on the observing conditions, since they are acquired with the camera shutter closed. Therefore, you don't need to take them each night during the observation run. You can make them once. For example, you can make darks during a night of bad weather (or even during the day, but beware of light leakage).

I encourage you to compile your own dark library (corresponding to *your* camera), varying the temperature (you can put the camera in the freezer to make it reach the low temperatures of the winter). Choose an exposure time longer than the longest exposure time you use for other images – e.g. 600 seconds. Take a large series of darks each time (say, thirty), and create "master darks" for each series.

Exposure Time

I suggest always using the same exposure time for your observations (typically 5 minutes – 300 seconds – for faint objects). Of course the total time of the acquisition will vary depending on the magnitude of the target, but you will control this using the number of exposure rather than their duration. This brings several advantages:

- You will be able to always use the same darks;

- Data reduction is faster (no parameter needs to be modified);

– There is no risk of making mistakes during the acquisition (since you don't modify the exposure time).

There are limitations to this method, especially for extremely faint objects. But in general, it is a trick to simplify your life.

Prepare your Observing Session

Much too often amateur observers wait until the last minute to decide what they want to observe. This rarely produces good results! Prepare your observations ahead, so that you can optimize the sequence of the targets based on their height in the sky, for example. You can even estimate the exposure time based on the magnitude of each target.

If you are looking for targets, or a theme for your observation, you can consult the Shelyak Instruments website[49] where we list several observing proposals for all levels.

Do not Wait to Reduce your Data

Generally, observers carry out the observation during the night, and wait until the next day(s) to reduce the data. Beware. If you wait too long, you run the risk of never reducing your data. And storing raw data on your drive is equivalent to doing nothing. I strongly recommend reducing the data right after the acquisition. You will still remember all the important information, and this reduces the chances of errors.

15.5   Share your Results

The best way of progressing fast is to compare your results with other observers. When you compare your data to those of a more experienced observer with a similar instrument, you can easily see how much room for improvement you have. For example, I encourage you to observe Be stars, and compare your results to those of the BeSS database, where you can probably find comparable observations. You can even share your own so they can be validated by an administrator, who will help you make improvements if needed.

You can also show your observations in the many forums available – for example the ARAS forum[50] or the Webastro forum[51]. Another way to share your observations is to set up your own website. The software VisualSpec provides robust functions to create HTML pages with spectra embedded.

---

[49] www.shelyak.com – spectroscopy tab / pedagogy & projects
[50] ARAS forum: http://www.spectro-aras.com/forum/index.php
[51] Webastro Forum: http://www.webastro.net/

Moreover, keep in mind that amateur spectroscopy is still in a pioneering phase. The data you collect are really interesting to the scientific community. Don't keep them for yourself.

15.6 Spectra of Professional Quality

Many amateur observers don't believe they can obtain spectra good enough to be shared or good enough to be compared to the results of professionals. This is a serious mistake. Most amateurs are perfectly able of getting physical measurements of great quality. Of course, amateurs do not have access to professionals quality equipment, but the quality of the data depends more on the observing protocol than on the instrument itself. A rigorous approach coupled with a lot of common sense will allow for results of professional quality. When you reach this level, you will be ready to participate actively to Pro-Am research programs. You can find more information on the Shelyak Instrument website. In a few words, to get high quality spectra, take special care in the following areas:

- precision of autoguiding

- reproducibility of the configuration and observing protocol;

- completeness of the data reduction: pre-processing, optimized extraction, calibration, correction of instrumental response, careful archiving;

- use a standard format (such as the BeSS format[52]);

- save and backup your raw data.

Never forget that, when you take a stellar spectrum, you are likely the only one doing it at that time. Do not keep this data for yourself, share it with the astronomical community.

---

[52] http://basebe.obspm.fr/basebe/

Conclusion

Here we are at the end of this book. I really hope that this is the start of a long adventure for you.

A proverb says "When the wise man points at the Moon, the idiot looks at the finger" (Confucius). I thought about this proverb many times while writing this book. These pages deal more with the finger (the spectroscopic equipment) than with the Moon (and more generally the sky). So am I an idiot?

To focus on the Moon (which is clearly the goal!) it is necessary to be able to forget about the finger. This means the motion of the finger as to be completely natural and automatic.

Obviously, spectroscopic observations require some effort. However, we are lucky to live in a time where the tools available (telescopes, mounts controlled via PC, CCD cameras, optical components, spectroscopes, software tools, databases, etc.) make our quest really accessible – taking good spectra is not reserved to a few adventurous DIY people.

It will take you some time, and dedication, to effectively control your instrument, and carry out spectroscopic observations as a daily routine. Only then you will be able to focus on the objects in the sky.

The game is worth the candle: by doing this initial technical effort, you can access a whole new level of astronomy.

This new level of achievement has several implications. Discovering the possibility of making a physical measurement on a star is a pleasure of its own. Discovering the variety of objects in the sky is another – and it is breathtaking. Little by little, you will understand better how researchers have built the astrophysics cathedral over time.

But there is even more. If you only follow the maintream media, you might feel like we understand everything in the sky, and that there is nothing more to discover. By carrying out spectroscopic observations, you will find quickly this is false. Even through your "humble measurements", you will soon encounter yet unexplained phenomena. Astrophysics is a living, sparkling science. The domains of active research are numerous: from cosmology, to stellar astrophysics, touching upon the search for extra-Terrestrial life.

And this is not all. In this sparkling science, there is a place for you! The amateur community has already provided brand new observations to researchers. The more amateurs contribute, the deeper the understanding that can be inferred from their data. So, on top of the new level of astronomy that you will discover, you will also see that amateur observers can contribute to the construction of human knowledge.

Your turn now!

François Cochard
Chamonix, July 2015

Glossary

The words defined here are underlined like this *in the text.*

Be star It is a particular type of star, targeted by both professional and amateur astronomers participating in the Pro-Am BeSS project, in collaboration with the Observatoire de Paris. They are characterized by having shown at least once a hydrogen Balmer line in emission. The emission lines can be stable or appear suddenly. The most common interpretation is that they come from excitation of a disk of hydrogen-rich material formed around the star. Be stars are thought to be massive, fast rotating, and barely stable. It can happen that large portions of their envelopes are ejected. They are a magnificent stellar laboratory. 15, 17-19, 45, 46, 62-64, 153, 228, 240

RA Right Ascension. It is one of the coordinates describing the position of the star in the sky (together with the declination – DEC). 195, 196, 206

ADU *Analog to Digital Unit* It is the unit of the intensity measure in a pixel. 129-131, 138-141, 158, 166, 173, 215

CCD Camera Camera that can acquire images in digital format 21, 115, 125-128, 132, 139, 164, 188

Continuum Part of the spectrum corresponding to the Planck Profile (or Blackbody profile). A spectrum is made of the continuum part on top of which emission and/or absorption lines are formed. 35, 166, 227, 228, 231

Blackbody Body that only absorbs light, without reflecting any. Thus, it is an isolated body, and it would be completely black if it were not emitting. The emission spectrum depends only on its temperature and follows the Planck profile. 32, 42, 98, 144

DEC Declination. It is one of the coordinates describing the position of the star in the sky (together with the right ascension – RA) 195, 196, 206

Raw image   It is the unprocessed output from the CCD camera. The values stored are necessarily integers, since they are the result of the conversion from analog to digital of the detector. 7, 100, 135, 153, 154, 173, 176, 177, 185, 221, 222, 230, 234, 238

Pro-Am   Collaboration project between amateur and professional astronomers. As spectroscopy is developing, so do these programs: do not hesitate to participate. 15, 62, 63, 241

1D Spectral Profile   It is a plot representing the spectrum, with in abscissa the wavelengths and on the ordinate axis the corresponding relative intensity. The 1D spectral profile is the final form of a spectrum that one gets at the end of the data reduction process. 8, 135, 153, 155, 159

2D Spectrum   Image of a spectrum, cf. ID spectral profile. The spectroscope can only produce 2D spectra, which are then turned into 1D spectral profile during data reduction. 8

References

References (books in english)

[1] James B. Kaler ; Stars and their Spectra [Cambridge University Press, 2011 second edition; ISBN: 978-0-521-89954-3]

[2] Marc F. M. Trypsteen et Richard Walker: Spectroscopy for Amateur Astronomers [Cambridge University Press]

[3] Richard Walker: Spectral Atlas for Amateur Astronomersn [Cambridge University Press]

[4] Ken M. Harrison: Astronomical Spectroscopy for Amateurs [Springer, 2011 first edition; ISBN: 978-1-4419-7238-5]

[5] John B. Hearnshaw: The analysis of starlight : two centuries of astronomical spectroscopy [Cambridge University Press, 2014 second edition; ISBN: 9781107031746]

[6] Richard O. Gray, Christopher J. Corbally: Stellar spectral classification [Princeton series in astrophysics, 2009 first edition; ISBN: 978-0-694-12510-7]

References (french books)

[1] Agnès Acker. *Astronomie Astrophysique - Introduction.*, 5e éd., 2013, Dunod.

[2] Jean Heyvaerts. *Astrophysique – Étoiles, Univers et relativité.*, 2e éd. 2012, Dunod.

[3] James Lequeux. *Naissance, évolution et mort des étoiles.* 2011, EDP.

[4] Bernard Maitte. *Une histoire de la lumière, de Platon au photon.* 2015, Seuil.

[5] Trinh Xuan Thuan. *La Mélodie secrète : Et l'homme créa l'univers.* 1991, Folio Essais.

[6] Trinh Xuan Thuan. *Le chaos et l'harmonie : La fabrication du Réel.* 2000, Folio Essais.

You can find a more complete bibliography on the Shelyak Instruments website (www.shelyak.com), with books in French and English.

Useful Links

I list here a few useful links.

Forums and mailing lists about spectroscopy

- **ARAS Forum** : Probably the most important amateur forum focused on spectroscopy.
  http://www.spectro-aras.com/forum/index.php

- **BAA Forum** : UK forum for discussed amateur astronomy with a section on spectroscopy. https://britastro.org/forum

- **Mailing-list spectro-L**. Main alert list (originally in french, now also in English) on spectroscopy. By subscribing, you will receive updates on all targets of opportunity and observations; you can also ask for help and/or advice.
  https://groups.yahoo.com/neo/groups/spectro-l/info

- **Mailing-list astronomical_spectroscopy**. Very active list (in english) on spectroscopy. Largest Internet site with over 1000 members. You can ask for help and/or advice.
  https://groups.yahoo.com/neo/groups/astronomical_spectroscopy/info

Personal webpages

- **Christian Buil website** : webpage full of information, observations, advice, and tests of equipment. A must.
  http://www.astrosurf.com/buil/

- **François Teyssier website** : François is very active in the community observing symbiotic stars and cataclysmic variables. His website has much information about these observations, and also many very useful tools (reference stars, proposed observations at low resolution, etc.).
  http://www.astronomie-amateur.fr/

Websites of associations

- **AAVSO** : American Association of Variable Stars Observers. The AAVSO is an international non-profit organization of variable star observers whose mission is to enable anyone, anywhere, to participate in scientific discovery through variable star astronomy.
  https://www.aavso.org/

- **AFA** : Association Française d'Astronomie. French national astronomy association, which spends a lot of effort to popularize astronomy.
  http://www.afanet.fr/

- **ARAS** : Astronomical Ring for Access to Spectroscopy. Informal association dedicated to spectroscopy. Useful complement to the eponymous forum. You will find tutorials, equipment tests, etc.
  http://www.astrosurf.com/aras/

- **BAA** : British Astronomical Association. The leading UK-based organization for amateur astronomers with an international membership. Operates a spectroscopic database. https://britastro.org/

- **CALA** : Club d'Astronomie de Lyon Ampère. One of the most important local associations (specifically, in Lyon) in France. Science popularization activities are held regularly, and often deal with spectroscopy.
  http://www.cala.asso.fr/

- **SAF** : Société Astronomique de France. Another large French national association, which as well as the popularization, also promotes collaborations between amateurs and professionals.
  http://lastronomie.com/

Software

- **ISIS** : *The* software for spectroscopic data reduction. Free, developed by Christian Buil.
  http://www.astrosurf.com/buil/isis/isis.htm

- **VisualSpec** : most widely used software used by amateurs all over the world to manipulate stellar spectra. Free, developed by Valérie Desnoux.

- **AudeLA** : Open source software for astronomy. It is a huge toolbox for all the needs of an observer: acquisition and image processing, telescope controls (autoguiding), dome control, spectroscopy. Can be customized with your own scripts. Free.
  http://audela.org/dokuwiki/doku.php?id=fr:start

- **Prism** : very complete piece of software for astronomy: acquisition and image processing, sky map, control of the instrumentation, scripts. Has built in functions for spectroscopy. Non-free, developed by Cyril Cavadore.
  http://www.prism-astro.com/fr/

- **MaximDL** : Another non-free very complete software.
  http://www.cyanogen.com/

- **Astroart** : Software for camera and telescope control and data analysis, includes autoguiding specifically for slit spectroscopes.
  http://www.msb-astroart.com/

- **PHD2** : Software for the autoguiding (now open source). Free.
  http://openphdguiding.org/

Websites of institutions

– **BeSS & ArasBeAm** : BeSS is the database of stellar spectra of Be stars, from both amateur and professional observers. It is the most complete catalog of Be stars. It is maintained and hosted by the l'Observatoire de Paris (LESIA). ArasBeAm is the complementary tool dedicated to amateur observers. This website indicates the stars to be observed with highest priority.
BeSS : http://basebe.obspm.fr/basebe/
ArasBeAm : http://arasbeam.free.fr/

– **CDS, Simbad** : Centre de Données Astronomiques de Strasbourg. It is *THE* main database for all astronomical data: catalogs, coordinates, images, spectra...the Simbad tool allows one to find easily the coordinates of any object in the sky. Essential.
CDS : http://cds.u-strasbg.fr/
Simbad : http://simbad.u-strasbg.fr/simbad/

– **Shelyak Instruments** : Company (of which I am the CEO) specialized in the design, manufacturing, and distribution of instruments dedicated to scientific astronomy, and specifically to spectroscopy. You will find, on top of the information on our products, several examples of observations, advice, etc.
http://www.shelyak.com/

– **OHP** : Observatoire de Haute-Provence. One of the top sites for astronomy in France. The 1.93 meter telescope made the discovery of the first exoplanet in 1995. Every year since 2004, we have organized a "Spectro Star Party", which attracts many amateurs and their instruments.
http://www.obs-hp.fr/welcome.shtml

Index

A

B

C

D

www.ingramcontent.com/pod-product-compliance
Lightning Source LLC
Chambersburg PA
CBHW042310210326
41598CB00041B/7332